THE PRENTICE HALL WORLD OF NATURE

the Environment and Conservation

MARTYN BRAMWELL

PRENTICE HALL

New York   London   Toronto   Sydney   Tokyo   Singapore

THE ENVIRONMENT AND
CONSERVATION

Managing Editor: Lionel Bender
Art Editor: Ben White
Text Editor: Alison Freegard
Assistant Editor: Madeleine Samuel
Project Editor: Graham Bateman
Production: Clive Sparling

Media conversion and typesetting
Peter MacDonald and Partners and
Brian Blackmore

AN ANDROMEDA BOOK

Devised and produced by:
Andromeda Oxford Limited
11–15 The Vineyard
Abingdon
Oxfordshire OX14 3PX
England

Prepared by Lionheart Books

Copyright © 1992 Andromeda
Oxford Ltd

All rights reserved. No part of this publication may be reproduced or utilized in any form or by any means, electronic or mechanical, including photocopying, recording, or by an information storage and retrieval system, without permission in writing from the publisher.

Library of Congress Catalog Card Number: 91-66996

ISBN 0-13-280090-X

Published in North America by:

Prentice Hall General Reference
15 Columbus Circle
New York, New York 10023

PRENTICE HALL and colophon are registered trademarks of Simon & Schuster, Inc.

Origination by Alpha Reprographics Ltd, Harefield, Middx, England
Manufactured in Singapore

10 9 8 7 6 5 4 3 2 1

First Prentice Hall Edition

Below: Sea anemones on rocky sea floor

CONTENTS

INTRODUCTION..............................5	POLLUTING THE WATERS 60
	POLLUTING THE LAND..................66
WHAT IS ECOLOGY? 6	POLLUTING THE AIR 70
FUELING THE SYSTEM 12	PESTS AND PEST CONTROL 76
FOOD CHAINS AND WEBS 18	SHARING PLANET EARTH.............80
THE NUMBERS GAME 24	WILDLIFE CONSERVATION 86
LIVING PARTNERSHIPS 28	
PLANTS AND SUCCESSION..........32	GLOSSARY 90
THE THREAT OF EXTINCTION 38	INDEX...92
THE PRESSURE OF PEOPLE.........48	FURTHER READING 94
DESTRUCTION OF FORESTS........56	ACKNOWLEDGMENTS...................95

INTRODUCTION

The environment is the surroundings of an organism and includes both the non-living world and the other organisms living in the area. No creature – plant, animal or micro-organism – that lives on Earth does so in isolation from other individuals or from its surroundings. Every living thing has an influence upon the others with which it comes into contact. And all creatures are affected by changes in the physical world – in the climate, landscape or seascape.

This book looks at how organisms interact with one another and with the non-living world. It examines the forces, pressures, cycles and systems within the environment. Some of these are natural, such as the effect of an increasing number of large meat-eating animals on populations of small plant-eating ones. Others are artificial or man-made. These include the effects of pollution on plant and animal communities. It also covers conservation and people's relationships with their fellow creatures.

The first section of the book describes the basic principles of ecology, that is how individuals of the same and different species of animals and plants interact with one another within communities of living things. Next it discusses how populations of organisms respond to and are affected by natural and artificial changes in their surroundings. This highlights the problems of a growing human population and the pressures this makes on the environment. A third section looks specifically at pollution of the environment and its consequences on the natural world. Lastly, the books deals with planet management – how we can look after the environment so that we and our fellow creatures continue to live in harmony.

Each article in this book is devoted to a specific aspect of the subject. The text starts with a short scene-setting story that highlights one or two of the topics described in the article. It then continues with details of the most interesting aspects, illustrating the discussion with specific examples.

Within the main text and photo captions in each article, the common or everyday names of animals and plants are used. For species illustrated in major artworks but not described elsewhere, the common and scientific (Latin) names of species are given in the caption accompanying the artwork. The index, which provides easy access to text and illustrations, is set out in alphabetical order of common names and of animal and plant groupings with the scientific names of species shown in parentheses.

A glossary provides definitions and short explanations of important technical terms used in the book. There is also a Further Reading list giving details of books for those who wish to take the subject further.

◄Clearing tropical rain forest in the Amazon not only destroys the home of the richest collection of wildlife in the world but may also upset the delicate balance of gases in the Earth's atmosphere.

WHAT IS ECOLOGY?

A lone male elephant wanders slowly through the wooded savannah of the Serengeti Plain in East Africa, stopping now and then to feed on foliage pulled from the branches of trees. This magnificent animal has no natural enemies except for human hunters, and its feeding methods enable it to share the plant-food resources of the savannah peacefully with other plant-eating residents including monkeys, giraffes, warthogs and antelopes.

No animal lives entirely on its own. It has a habitat, its natural home, which may be forest, desert, the deep ocean or the bleak summit of a mountain peak. It breathes air and it drinks water. It must feed itself and in turn it may be hunted by meat-eaters. These links between individual animals, and between animals and their surroundings, make up the science of ecology.

THE SCIENCE OF ECOLOGY

Studying any individual animal on its own will provide only a very limited amount of information. We may find out how it breathes and moves, and how it reproduces, but to really understand how it lives, and how – and why – it behaves in a particular way, we must also look at where it lives, who else lives there, who eats what, and how the different animals in that habitat are equipped for the conditions there. To understand all this, it is useful first to know a little about how life evolved on Earth.

THE BEGINNINGS OF LIFE

The Earth was formed about 4,600 million years ago. It is known from some of the oldest rocks that water was present on the surface only a few hundred million years later. However, at this time there was still no oxygen in

► Ecology, the study of animals and plants in relation to their environment, can be approached at many different levels. One scientist may choose to study individual animals or plants. Another may study large groups, or populations, of a single species. Taking a wider view, another may prefer to study a varied community of different animals and plants living close together – for example in a hedgerow or an oak tree. A wider view still would take in the whole of an ecosystem, such as a woodland, marsh or lake. Each approach will tackle different questions and require slightly different methods.

▼ Deciduous trees are perfectly adapted to cope with the changing seasons. Most of their growth takes place in spring and summer when they also produce their fruits and seeds. In the fall, the trees shed their leaves and slow down their biological processes, conserving their energy through the winter months.

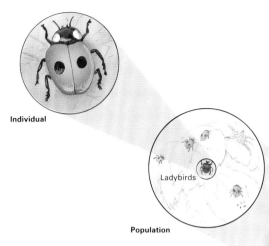

the air: the atmosphere was a lethal mixture of poisonous gases and water vapor. There was no life on land, and nothing we would recognize as animal life in the seas. In this barren state the primitive Earth awaited the appearance of the very first living cells.

The earliest forms of life so far discovered are microscopic single-celled organisms and a few simple algae (members of the seaweed family) that appeared in the Earth's ancient seas. Fossil traces of these organisms have been found in rocks over 3,300 million years old.

BUILDING WITH LIGHT

At some point in the distant past, a very important change occurred. Some of these primitive cells started producing a green chemical called chlorophyll. Using this the cells were able to manufacture their own food from water and carbon dioxide, using the light of the Sun as a source of energy. These cells were the very first form of plant life, and the remarkable process they had developed is called photosynthesis ("building with light"). Its main products are sugars and starches – the chemical building blocks from which all plants are made. Its all-important by-product is the life-supporting gas oxygen.

At first, most of the oxygen released by these primitive plants was used up in chemical reactions that produced new kinds of rocks and minerals, but soon there was oxygen to spare – and the gas began to build up in the atmosphere. At this point the scene was set for animal and plant evolution to begin in earnest.

EVOLUTION TAKES OFF

At first, progress was still very slow. Like the first plants, the earliest animals were also very simple single-celled organisms, probably very like the ones that are common in sea and pond water today. The pace soon quickened, however, and by 600 million years ago the primitive seas contained a wide variety of animals including sponges, shellfish, sea snails and trilobites – ancient relatives of the modern crabs and spiders.

Plants moved out of the seas and on to dry land about 400 million years ago. The first ones were probably little more than modified seaweeds, but true land plants soon evolved, with strong upright stems to support them. Soon after this the first land animals appeared. In the millions of years since then, thousands of different species have evolved. Some survived only briefly; others for enormous periods of time. The great reptiles of the dinosaur families, for example, dominated life on Earth for over 140 million years. Humans, the relative newcomers of the animal kingdom, have been around for only a few

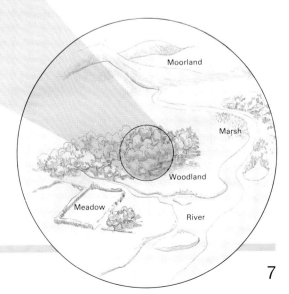

million years but in that short time have evolved far beyond any other form of life in the history of the Earth.

TRIAL, ERROR AND SURVIVAL

Every living organism carries a plan, a chemically coded "blueprint for life" in each of the cells in its body. The code is carried on pairs of chromosomes (23 in human beings), which are spirally coiled ribbons of a special molecule called DNA. Along each chromosome there are individual pieces of code called genes. It is the genes, in various combinations, that determine an animal's size and shape, the color of its fur or feathers, how good its sense of smell or sight will be, and how fast it runs, flies or swims.

Every animal gets half its genes from one parent and half from the other. As each new generation is produced, these bits of code are shuffled like a pack of cards to give endless different combinations and variations. (That is why brothers and sisters may share a family likeness but are never exactly like either of their parents). These variations can even turn out to be like living experiments. An animal that is better than the rest at finding food, or at finding a mate, or one that is good at avoiding its enemies, will survive – and will then breed and pass on those "useful" genes to its offspring. Other individuals that are not as well equipped will fail to raise healthy young, and their family lines will eventually die out.

This is the basis of evolution, first described by the famous British naturalist Charles Darwin in 1859 in his book, *On the Origin of Species by Means of Natural Selection*. We sometimes refer to it as "the survival of the fittest," using "fittest" to mean not just the animal that is the healthiest, but the one that is best suited to its environment.

▲ Sea urchins, scallops, crabs and seaweeds make up a large part of the seabed community in this Scottish loch. The crabs are predator/scavengers, the urchins mainly eat algae, and the scallop filters food from the water.

▼ These diatoms are microscopic plants which drift in the surface waters of the oceans. Like land plants they can make food using the Sun's energy, and just like the savannahs on land they provide food for plant-eating marine animals.

ECOLOGY TODAY

Biologists and environmentalists today need to have a wide range of scientific skills. They must study individual animals, large groups (or populations) of a particular species, and mixed groups of animals and plants all living in the same habitat. Also, increasingly important in the modern world, they must be equipped to study the impact on the natural world of the rapidly growing human population and its demands for food and fuel, raw materials, living space and land for agriculture.

ANIMAL POPULATIONS

A group of animals of the same species is called a population, and for many animals, especially the social ones (those that live together in groups, like deer, bees, dolphins and rabbits) the population is a very important unit.

The local population at any particular place will vary according to the number of births and deaths there are, and whether the animals stay in one place or migrate from place to place. These factors in turn depend on weather conditions, and especially on how good the food supply is. A mild winter and an early spring will usually produce large populations of insects. As a result of this abundant food supply, and an early start to the breeding season, insect-eating birds such as tits and warblers will produce large families, so their population too will be large. After a harsh winter there will be far fewer insects available, and many insect-eaters will respond by breeding later and by producing only small families.

ANIMAL COMMUNITIES

Many different animals and plants living close together make up a community. Some will be plant-eaters, some will be hunters. Most will move around independently, feeding, resting and raising their young, while others, called parasites, hitch a ride on someone else.

Plant parasites such as mistletoe grow on another tree, called a "host," and send their roots into its branches to "steal" a food supply. There are animal parasites too, such as the fleas and lice that live on birds and mammals, and the tapeworms and other internal parasites that live inside the host's body. Most of these parasites live on blood, but although some eventually kill the host, many others strike a balance – taking what they need without damaging the host too much. In survival terms this appears to be a much more efficient system.

▼ In dry regions, the natural vegetation is easily overgrazed if there are too many animals. Here, in Australia, the grasses and shrubs are almost gone and the soil has been trampled hard.

A parasite that kills its host must very quickly find a new one, or perish.

There are also many other kinds of relationships, between animals of different kinds, and between plants and animals, where neither does any harm at all to the other. Cattle egrets, for example, ride around on the backs of wildebeest or domestic cows, feeding on the insects stirred up by the animals' feet as they graze. In return, the birds provide an early warning system which alerts the grazers when lions or leopards are on the hunt. Other relationships are even closer. As described later, many plants rely on animals to pollinate their flowers, and they attract these animal helpers with enticing sugar-rich nectar.

ECOSYSTEMS

The most complex units studied by ecologists are ecosystems, which were given their name in 1935 by the British scientist Sir Arthur Tansley. The ecosystem goes quite a lot further than the community because it includes not only the plants and animals but also the physical and chemical environment in which they live. Size is not particularly important: what does matter is that the system works as a complete unit, with all the light, water, minerals and other nutrients necessary to keep its plant and animal communities alive and in good health.

An ecosystem can be quite small, for example a pond, a seaside rockpool, or a decaying log on the forest floor, but it can also be extremely large, such as the Arctic Ocean, the Amazon rain forest, the Sahara Desert or the North American prairies.

▶The gorilla, largest of all the apes, is a gentle giant whose only real enemy is man. An adult male may weigh 350lb, (average human male 160lb), his mate about 200lb. The animals feed almost entirely on soft young leaves, and the family group stays close together, seldom moving more than 1mi in the course of a day.

FUELING THE SYSTEM

As dawn breaks over the French countryside the bright yellow flowers of a celandine plant open and turn to face the Sun. For the rest of the day, like a tiny radar dish, each flower will track the Sun across the sky while the plant's leaves soak up the Sun's energy and use it to manufacture food. Towards evening the celandine is eaten by a rabbit, but its job is done. Energy stored in its leaves has been passed to the rabbit, while food-energy stored in its underground tuber will fuel the growth of a new celandine next spring.

Every living thing on Earth needs food in some form or other. The food provides two things: raw materials for building new plant or animal tissues, and fuel for energy to drive the building processes. Every animal on Earth depends on plants in order to live. Some, such as elephants, caterpillars, finches and rabbits, feed directly on plants. Others, such as lions and weasels, piranha fish and hawks, feed on other animals – but the animals they hunt are nearly always plant-eaters, or occasionally they feed on other, smaller hunters that in turn prey on plant-eaters.

HOW FOOD IS USED

Animals fuel their systems by taking in food and breaking it down into its chemical building blocks by the process of digestion. Some of the chemicals are used to build new bone cells, muscle cells, hair, fat or blood cells. The rest is "burned" to provide the energy necessary for moving about, catching food, digesting it, building new cells, and keeping the body at working temperature.

Plants function very differently. Unlike animals they can manufacture the food they need, and because they do not chase about, or need to keep their bodies warm all the time, they can produce all the materials and energy they need from water, dissolved nutrients, carbon dioxide and sunlight.

HARNESSING THE SUN

The plant's chemical factories are in its green parts, usually the leaves. Many plants have green cells in the stem as well, however. These parts of the plant owe their characteristic color to the green chemical chlorophyll which is held in special structures called chloroplasts inside the leaf cells. It is here that the remarkable process of photosynthesis takes place.

Chlorophyll has the unique ability to absorb energy from sunlight and use it to break down water molecules into their components, hydrogen and oxygen. Some of the oxygen escapes through tiny holes in the leaf surface, and this helps to keep up the oxygen level in the air we breathe. The

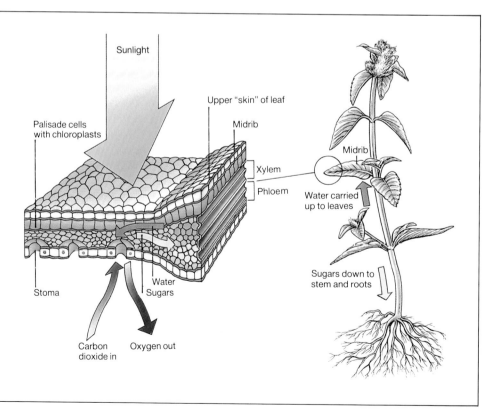

▶ **Photosynthesis**
Photosynthesis is one of the most important chemical processes ever to evolve on Earth. It converted Earth's original poisonous atmosphere into one that could support life. All the time it refreshes the atmosphere by using up carbon dioxide and adding new oxygen to the air around us.

The process takes place inside the plant cells, in tiny green "factories" called chloroplasts. These get their characteristic color from the pigment chlorophyll, which acts as a catalyst – a chemical that enables a particular chemical reaction to take place. Cells under the leaf surface are packed with chloroplasts, and here the Sun's energy is converted into chemical energy. This energy is then used to combine carbon dioxide and water to make sugar and oxygen. Some of the sugar is used by the plant as fuel to drive other chemical processes, while some is converted to starch and stored.

hydrogen stays inside and combines with carbon dioxide from the air to build up molecules of sugar, the main building blocks of plant cells.

DAY SHIFT – NIGHT SHIFT

Although plants do give off oxygen, they do not give off all the oxygen they produce. Like all living things they need oxygen in order to live. They do not breathe like animals but they need oxygen in just the same way for the chemical processes that keep them alive. What happens is that the plant has a day shift and a night shift.

During the day, photosynthesis is fuelled by the Sun. The green cells are hard at work making food and building materials, and the oxygen the plant needs for its own use: only the spare oxygen is released into the air. At night, there is no photosynthesis. The chloroplasts shut down during the hours of darkness and the plant continues to function by taking in the oxygen it needs from the air and giving off carbon dioxide.

Plants can only use up large amounts of carbon dioxide and give off large amounts of oxygen when photosynthesis is running at top speed, and this is one reason why the cutting down of the tropical forests is causing such concern. It is these huge

▼ The Sun's energy is "fixed" by plants and converted into usable food, but some ecosystems are more efficient than others. Tropical forests are the most efficient, followed by temperate forests and grasslands. Oceans are less efficient but because of their huge size still make an important contribution.

▲ Although green vegetation all over the world contributes to the health of the atmosphere, no other habitat can match the chemical productivity of the tropical rain forests. The constant sunlight, warmth and moisture here on the island of Dominica and Haiti allow a very high rate of photosynthesis all year round.

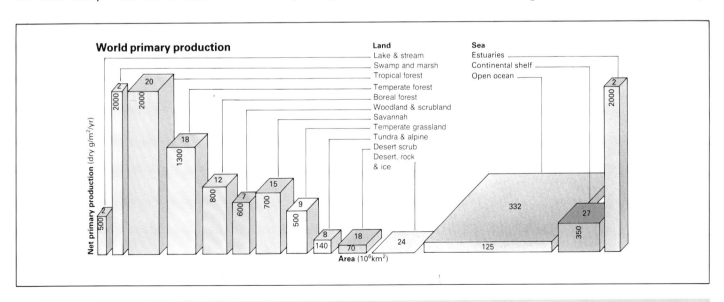

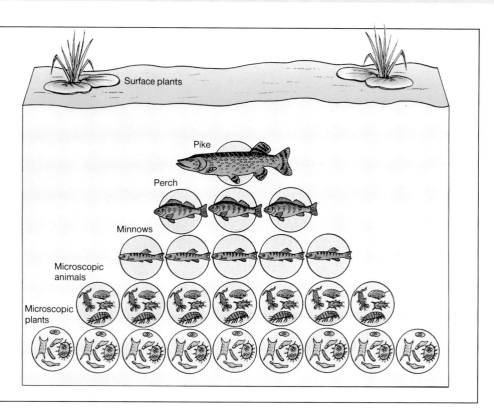

▶ **The food pyramid**
The pyramid-like arrangement of feeding relationships in a freshwater pond is very like that on land. At the base of the pyramid are the plants, which harness the Sun's energy and convert it into food which can be used by the animals higher up the pyramid. There are large water plants that float on the surface with roots trailing in the water, or that have their roots in the mud at the bottom, but a large part of the photosynthesis is carried out by microscopic plants that float in the surface waters. The plants are the energy fixers, and this level is therefore called the producer level.

Tiny floating and swimming animals such as water fleas and insect larvae form the first consumer level, and they in turn provide food for small fishes. Above them are the predators, and right at the top of the pyramid is the chief predator, which usually has no natural enemy apart from humans.

areas of leafy forest in the sunniest parts of the world that keep our atmosphere healthy. Their loss would rob the Earth of its main biological powerhouse.

PLANT-EATING ANIMALS
One problem with eating plants such as leaves and grass is that they are difficult to eat and difficult to digest. The cell walls are made of cellulose, which is very tough, and so most leaf- and grass-eaters have sharp chopping teeth at the front of the mouth and large, flat, ridged teeth at the back to crush the food and grind it to a pulp. Sheep, cattle, elephants and horses all have this kind of equipment.

The stomachs of many grass-eaters are also special. Sheep and cows, for example, are ruminants – animals that "chew the cud." Their stomachs have several separate parts or chambers. Food passes first into one part of the stomach where acids and special bacteria help to break it down. The animal then "coughs it up" and chews it again. Finally it passes into another part of the stomach and then into the intestine for further digestion.

Processing food this way has several advantages. Leaves and grasses are low in food value, which means the animal has to eat a huge amount in order to keep itself alive. The ruminant system means that the food remains in the stomach for a long time, which gives the acids and bacteria plenty of time to work.

Having to eat huge amounts of low-quality food also means that many plant-eaters are large, slow-moving creatures. Grassland animals, for example, are always in danger from hunters, and this method of feeding allows them to eat a large amount of food fairly quickly while in the open, and then retire to the relative safety of woodland or long grass while they digest the meal.

THE MEAT-EATERS
Meat-eaters, although they have to catch all their food, do not have the plant-eaters' problems of having to process large quantities of low-quality food. Animal flesh, insects, fish and birds' eggs are all high in food value and easily digested. Carnivores are therefore often smaller than herbivores – and built for speed and power.

Their teeth are very different too. Many have sharp stabbing fangs with which to seize and kill their prey. Knife-like shearing teeth at the side of the mouth will cut through tough hide or sinews, and flat-topped teeth at the back are used for crushing bones. Different types of hunters have different combinations of teeth, each one suited to the kind of food the animal eats.

▶ Large herbivores are less common in forests than in grasslands, but Red deer and other species do occur in forests, especially where the forest cover is broken up by grassy glades. It is now known that Stone Age people cleared areas of forest to encourage these game animals, as a source of meat, and for their antlers and hides.

THE ENERGY PYRAMID

It soon becomes clear that in any patch of woodland there are huge numbers of tiny creatures, especially insects, feeding on plants. There are not nearly so many medium-sized animals, such as birds, mice, squirrels and deer, and there are very few of the top hunters such as owls, hawks and foxes. The same general pattern is true whether it is in the tropical rain forest, the ocean or the African savannah. The main reason for this lies in the way energy is passed upward from one level to another through the system. When an animal feeds, only a small proportion of the energy it takes in is used to make it grow bigger. In fact once it is fully mature it will not grow at all but will simply be repairing and maintaining its body. Most of the energy it takes in as food is used to drive the living processes – feeding, breathing, walking or flying, digesting food, and so on. As this energy is burned as fuel for the muscles and chemical processes in the body, it finally escapes into the atmosphere as heat. That energy is lost: only the energy stored in the animal's body is available as food-energy for another animal to use. So, it takes a huge number of insects to feed just one bird, and many small birds to support just one owl or hawk.

Some idea of the numbers involved can be seen from studies of nesting swallows. In a single day, a pair of Barn swallows with a nest full of chicks

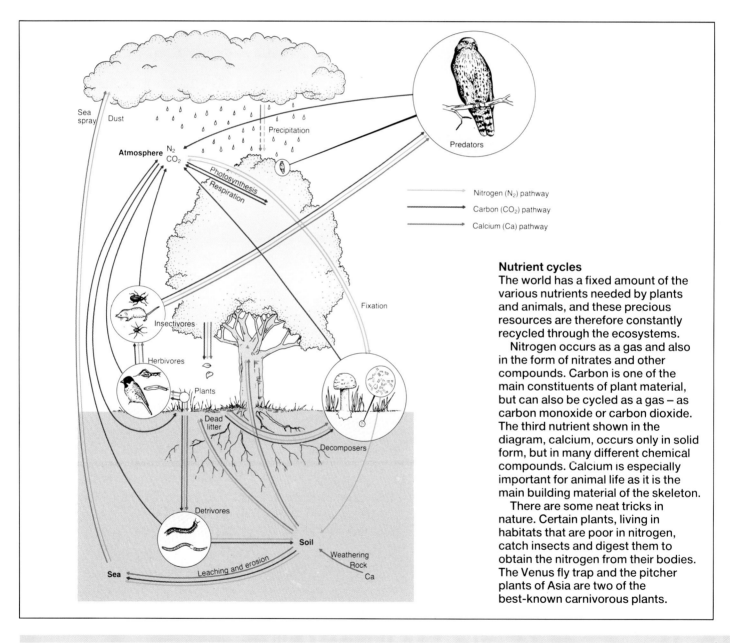

Nutrient cycles
The world has a fixed amount of the various nutrients needed by plants and animals, and these precious resources are therefore constantly recycled through the ecosystems.
  Nitrogen occurs as a gas and also in the form of nitrates and other compounds. Carbon is one of the main constituents of plant material, but can also be cycled as a gas – as carbon monoxide or carbon dioxide. The third nutrient shown in the diagram, calcium, occurs only in solid form, but in many different chemical compounds. Calcium is especially important for animal life as it is the main building material of the skeleton.
  There are some neat tricks in nature. Certain plants, living in habitats that are poor in nitrogen, catch insects and digest them to obtain the nitrogen from their bodies. The Venus fly trap and the pitcher plants of Asia are two of the best-known carnivorous plants.

▲ The nesting islands of many seabirds, like these Brown pelicans off the coast of Peru, are covered in thick deposits of bird droppings, or "guano." The droppings are rich in phosphorus, and many of these islands are mined to provide fertilizer for agricultural use.

made over 400 hunting trips between them, catching more than 8,000 insects to feed their hungry brood.

The system remains in balance only because the total energy supply is topped up constantly by plants harnessing the Sun's energy.

OTHER RAW MATERIALS

The energy supply from the Sun is constantly renewed, but the same is not true of the other raw materials necessary for life. In order to build their stems and leaves, or muscles, bones and hair, living things need many different raw materials. The most important in terms of quantity are carbon, hydrogen and oxygen, but plants and animals also require small amounts of nitrogen, potassium, calcium, sulfur, phosphorus and many other elements.

Plants take in all these nutrients through their roots, dissolved in the water in the soil. Animals obtain them from the plants and animals they eat. Plants, for example, need small amounts of calcium to make their cells function efficiently, but animals need much more of this chemical because it is the main building material for bones, teeth and shells.

In areas where the soil contains a good supply of calcium, there are usually plenty of snails. In acid heathlands, which are poor in calcium, there are hardly any snails, but slugs are quite common. They are close relatives of the snails – but of course have no shells.

RECYCLING THE NUTRIENTS

The main difference between energy and nutrient supplies is that nutrient supplies are not replaced, unlike the constant supply of energy from the Sun. The Earth contains all the calcium, potassium and so on that it will ever have. There is no new source of supply, and so these nutrients have to be constantly reused. The story of how these essential nutrients are recycled begins in the soil.

THE HIDDEN HEROES

The workers, unsung heroes of the animal world, are the tiny organisms that live in the soil, and in the layer of dead leaves, twigs and animal droppings that lies on the soil surface. They are the decomposers, and without them the Earth would long ago have been buried under a thick layer of organic debris.

The decomposers vary in size from microscopic bacteria, through tiny insects no bigger than a pin-head, to larger animals such as earthworms, ants and the adults and larvae of countless bugs and beetles.

From the moment a bird or mouse dies and falls to the ground, the decomposers take over. Bacteria start to break down the remains into their chemical building blocks. Scavenging animals such as Carrion crows, magpies, badgers and foxes feed on the remains and help to break up the body, and carrion beetles lay their eggs in it so that their grubs hatch out in the middle of a larder full of food.

Leaves on the forest floor rot away into a soggy mass of compost, busy with woodlice, millipedes, mites and worms. One study of an English beechwood revealed a population of 880 million mites in a single acre of forest floor litter. Fallen branches are soon riddled with the boreholes of bark beetles and other insects whose grubs feed on decaying wood.

Perhaps most important of all in breaking down branches and tree trunks are the fungi. These strange organisms are sometimes grouped with the plants, but unlike true plants they contain no chlorophyll.

WOODLAND KILLER

Fungi cannot make their own food and must therefore steal it. Some are saprophytes, which live entirely on dead organic material. Others are parasites, taking their food from living trees, and often killing the trees in the process. The Honey fungus is a common example. Its tawny-brown toadstools appear in clusters around the bases of dead or dying trees. The toadstools are the reproductive parts of the fungus but the main part is inside the tree. It consists of a mass of fine threads, like cotton wool, that penetrate deep into the living wood, absorbing nutrients and eventually killing the tree.

FOOD CHAINS AND WEBS

It is summer in the arctic tundra, and the landscape is carpeted with grasses, mosses and fungi taking advantage of the few brief months when the land is not covered in ice and snow. Crouched behind a rock an Arctic fox is watching a ptarmigan feeding. Suddenly – the fox dashes from cover. The bird gives a harsh cry of alarm and frantically beats its wings, but with an agile leap the fox catches it just as it takes off.

That scene in the Arctic must have been seen many times over during the 1920s by a young British biologist called Charles Elton. At that time he was studying at the University of Oxford, and during several expeditions to Bear Island, near Spitzbergen, he became fascinated by the links between the animals and plants of the arctic tundra. Other biologists were studying animals and plants in many different habitats, but the Arctic regions had one very special attraction. Because there are no trees to give cover, even the largest animals can be watched quite easily.

LINKS IN THE CHAIN

Elton soon became curious about who was eating what. The fox is one of the main hunters of the arctic tundra, and its main prey are birds such as ptarmigan, Snow buntings and sandpipers. Elton then observed what these birds were eating, and found that the ptarmigan fed on the berries and leaves of tundra plants while the buntings and sandpipers fed mainly on the plant-eating insects and insect larvae that are so abundant in the tundra summer. Elton had found two clear "food chains." One had three links: tundra plants–ptarmigan–fox. The other food chain had four links: tundra plants–insects–Snow bunting/Purple sandpiper–Arctic fox.

When Elton looked at the other main hunters of the arctic tundra, the Snowy owl and the Rough-legged buzzard, he found that they too were the final links in similar chains. The buzzard, for example, preys mainly on Arctic lemmings and Tundra voles – small rodents whose main food is the roots and shoots of tundra plants.

Elton's ideas about food chains were an important step forward in the new science of ecology. They led on to the "pyramid of numbers" theory (see page 14) and to an understanding of how the size of an animal is directly linked to the food it eats. They also helped to explain the distribution of animals of different sizes. The main hunters are not only big, and rare, but also need large territories. A male tiger, for example, needs several square miles, while a large garden is enough space for a robin.

The decomposers
The leaf litter covering the floor of a wood or a large garden is a hive of industry. Huge numbers of small creatures feed on dead and decaying plant and animal remains, making the nutrients available for re-use by new plants. Among the most common decomposers are cockroaches (*Ectobius lapponicus*) (**1**), slugs (*Limax maximus*) (**2**), Black-lipped hedge snails (*Cepaea nemoralis*) (**3**), woodland millipedes (*Cylindroiulus* species) (**4**), False scorpions (*Dendrochenes cyrneus*) (**5**), earthworms (*Allolobophora turgida*) (**6**), termites (*Reticuli fermes*) (**7**), centipedes (*Lithobius forficatus*) (**8**), sexton beetles (*Necrophorus* species) (**9**), hoverflies (*Chrysotoxum cautum*) (**10**), Flat-backed millipedes (*Polydesmus complanatus*) (**11**), and assorted fungi (**12**).

GARDEN FOOD CHAINS

To find examples of food chains it is not necessary to travel to the far north, or the jungles of India. They are all around us in the country, towns and cities, in parks and gardens – even in a window box of an apartment block.

Look at an ordinary garden in summer. Caterpillars and snails will be feeding on some of the leaves while greenfly suck up all the juices from the gardener's favorite rose stems. A Song thrush may swoop down on one of the snails, break open its shell on a nearby stone, and then eat the animal inside. Later in the day the thrush may be caught by a sparrowhawk, visiting the garden from nearby woodland.

In another part of the garden a ladybird is preying on greenfly – but then the ladybird flies into the web of a spider and is killed. Moments later the spider ends up as a tasty morsel for a Blue tit – which finally falls prey to the neighborhood cat.

These food chains may not seem as exciting as those of the African savannah or the South American rain

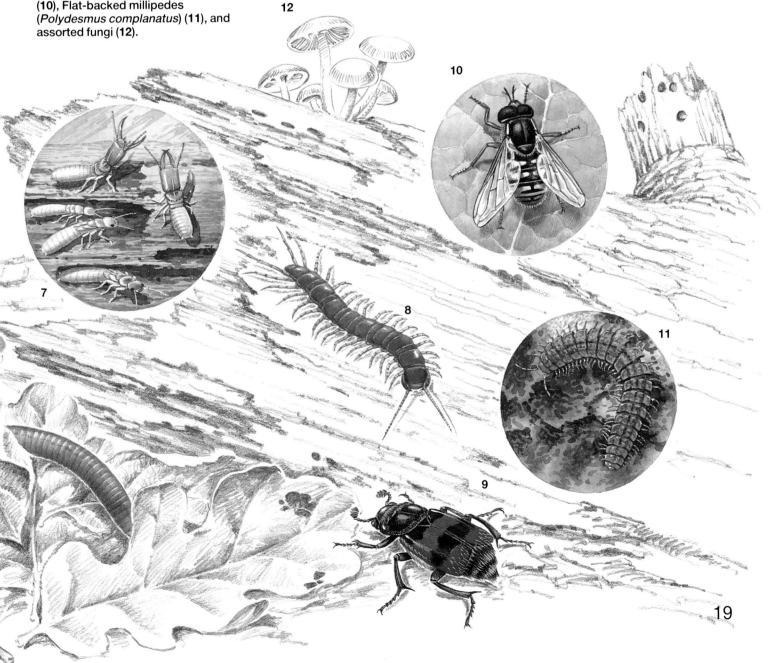

forest. However, to the biologist and ecologist they are exactly the same.

WHAT IS A FOOD WEB?

The idea of a food chain is simple enough, and it is a very useful tool for the scientist when working out how energy is passed along from one animal to another in an ecosystem. However, the real world is a little more complicated.

The Blue tit mentioned earlier does not feed exclusively on spiders. It catches insects by the thousand, and will take flies, grubs, caterpillars, ants and anything else it can catch. The cat will also take a variety of food. It is a carnivore, but it does not only catch birds: if the chance comes along it is just as likely to pounce on a mouse, a young rabbit, a lizard or fish, or even a large insect such as a grasshopper.

Some animals are even less specialized and will take a mixture of both plant and animal food. The hedgehog is a good example. It feeds mainly on earthworms, beetles, slugs and caterpillars, which it finds by rooting amongst the leaf litter with its sensitive nose. However, it will also scavenge the remains of any dead animal it finds, take birds' eggs and nestlings if it can reach them, and also feed on berries, seeds and fallen fruit. Even the Snow bunting, which Charles Elton watched feeding on insects in the arctic tundra, includes seeds in its diet.

From this it is clear that a series of simple "A eats B eats C" chains cannot possibly give an accurate picture of the world of the Blue tit or hedgehog. Most food chains have many side-branches representing the alternative foods an animal can take, and as these side-links cross over each other they become linked together to form what is called a food web.

PRODUCERS AND CONSUMERS

In the previous section "The Energy Pyramid" (page 16) described how energy was passed upward through an ecosystem from the plants that originally harnessed the Sun's energy, through the plant-eaters and on to the flesh-eaters. Ecologists call these different stages trophic levels, which simply means "feeding levels" (from the Greek word *trophikos* meaning "nourishment").

Food web in a temperate lake

○ First trophic level (primary producers)
○ Second trophic level (herbivores)
○ Third trophic level
○ Fourth trophic level
○ Fifth trophic level
○ Sixth trophic level

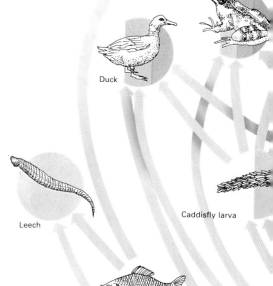

◀▲**Food webs and food pyramids**
The diagram on the left shows plants and animals arranged in a series of trophic (feeding) levels. Here, a plant community of three species acts as the energy fixer, supporting 12 plant-eating species and three levels of carnivores.

The main diagram above shows just some of the feeding links in a freshwater lake. Water plants in the shallows and microscopic phytoplankton in the surface waters fix the energy of the Sun and provide food for numerous herbivores. The web of feeding links then spreads upwards through several levels of carnivores until it reaches the egret.

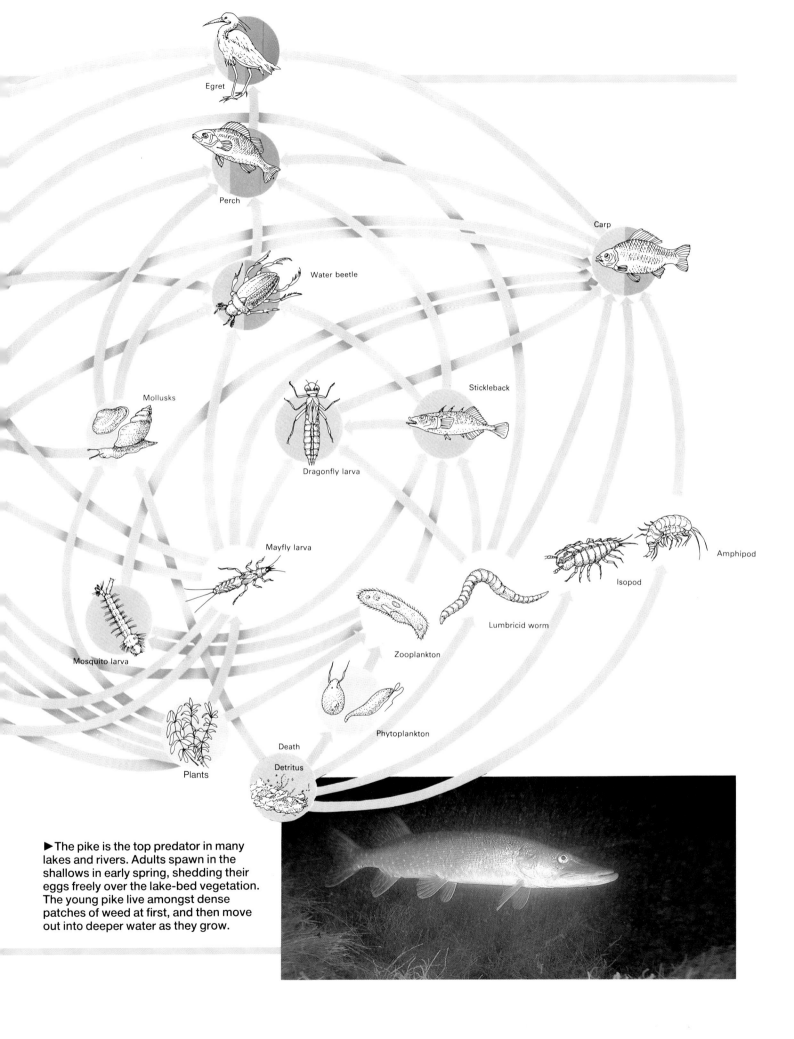

►The pike is the top predator in many lakes and rivers. Adults spawn in the shallows in early spring, shedding their eggs freely over the lake-bed vegetation. The young pike live amongst dense patches of weed at first, and then move out into deeper water as they grow.

A QUIET LIFE AT THE TOP?

The first trophic level consists of the plants, often referred to as the "fixers" or "producers" because they capture the Sun's energy and turn it into food which can then be used by the animals at higher levels.

The second level consists of the plant-eaters or herbivores – a huge group of animals ranging in size from tiny insects, through rabbits and mice to large grazing and browsing animals such as deer, antelopes, buffalo and elephants. These animals are the first level of the "consumers" – the animals that make use of the food energy fixed by the plants.

Above the plant-eaters are the first rank of carnivores, or flesh-eaters. They too are a large and varied group, ranging from predatory insects like the praying mantis, to shrews and a host of insect-eating birds.

Finally, right at the peak of the energy pyramid, there are the top predators – the hawks and eagles, killer whales and sharks, and the big cats such as the lion, tiger and leopard.

In the natural world these animals have no enemies. Left in peace they would die only from old age, or perhaps from an injury or disease. Unfortunately, they are not left in peace. They have one dangerous enemy – humans – and it is a fact that many of these superb creatures are now on the long list of Endangered Species. Several have been hunted almost to extinction for their beautiful fur or feathers, some have been branded as sheep-killers (often quite wrongly) and have been persecuted by farmers, while many others are threatened simply because their habitats are being destroyed.

A JOB FOR LIFE

Each species of plant and animal occupies a special place in the natural world, depending on where it lives, and how it lives its life. This special place is called its "niche," and it is rather like having a job description. For example, in old established woodlands there are many dead trees, broken branches and trees with dying limbs, all riddled with wood-boring insects and their grubs. That situation provides a perfect position or job for an animal that can cling securely to the side of a tree trunk, hack an opening in the rotting wood, and extract the insects and grubs from their maze of boreholes. Evolution has come up with the perfect candidate – the woodpecker.

A woodpecker's feet have two toes facing forward and two back to provide a firm grip on vertical surfaces; its tail feathers are very stiff and can be used as a prop to support the body; it is equipped with a powerful chisel-like bill, and neck muscles that act like shock absorbers so that it can hammer its way into decaying wood; and it has an enormously long sticky tongue with which to reach in and pull out its food.

SHARE – OR DISAPPEAR?

Some animals are even more specialized than the woodpecker, others not quite so specialized, but there is never more than one animal filling exactly the same niche in the same place at the same time. However, "time-share" systems can work very well. The Barn owl and the kestrel both feed mainly on mice, and in very similar habitats, but the owl feeds almost entirely at night while the kestrel hunts by day. That way the two birds can overlap without having to compete with each other for the same food. The niches of shrews and moles also overlap, but in a rather different way. Both animals eat earthworms, which live in the surface soil and leaf litter, but the shrew feeds on the surface while the mole burrows underground, so once again the two are not competing.

When two animals do try to occupy the same niche there is a problem, and one will usually come out on top. In New Hampshire in the United States, up to about 130 years ago the American cabbage butterfly was widespread in woodlands and hayfields. Then, in the 1860s, the European species was introduced and numbers of the native

▼ A Brown rat ends up as a meal for an Australian carpet python, beautifully patterned to blend with the forest floor leaf litter. The snake is able to swallow prey larger than its head by temporarily "uncoupling" the bones of its jaw.

▲ The Peregrine falcon is the supreme aerial hunter of the north temperate region. It feeds mainly on birds such as the Golden plover seen here, usually killing its prey with a blow from its talons after a near-vertical 130mph hunting dive known as a stoop.

species soon declined. The caterpillars of the European species could feed on a wider variety of plants.

NATURE'S ALL-ROUNDERS
In addition to a bewildering array of specialists, the natural world also has its generalists. These animals are called omnivores, and they take a mixture of animal and plant food. Most primates, including humans, belong to this group. Baboons, for example, are primarily fruit-eaters but they also eat flowers, bark, roots and insects, and will kill small birds and other mammals when the opportunity arises. Other omnivores include pigs and rats, which is why these animals can have such a damaging impact on island ecosystems.

ACCIDENTAL INVADERS
The idea of the niche helps to explain why disastrous events can follow the introduction of predatory animals into a new habitat, especially an isolated one such as an island.

Many Pacific and Indian Ocean islands, for example, have no native mammal predators. As a result they became the home of large numbers of birds, many of which nest on the ground or in low bushes. In the great age of ocean trading, sailing ships called at many of these islands to take on fresh water and collect fresh food. Rats escaped from the ships and took up residence ashore, where they had no natural enemies but a huge supply of easily caught food in the form of birds' eggs and nestlings. The rats had found an unoccupied niche, and they took it over immediately. Many native birds were almost wiped out.

Similar disasters followed when cats escaped from ships, and when pigs and goats were deliberately set free on mid-ocean islands to fend for themselves, breed, and so provide a source of food when the ships called again.

THE NUMBERS GAME

As rainclouds sweep across the Serengeti Plain of East Africa, bringing an end to the long dry season, millions of wildebeest, zebra and gazelles gather in herds that stretch as far as the horizon. The wildebeest young will be born as the herds migrate south. Many will be killed by lions and hyenas, others will perish in swollen rivers or from exhaustion, but enough will survive to keep up the strength of the herd, and to breed and have their own young in future years.

The herds of Serengeti grazers spend a large part of each year on the move. The main wet season, January to May, is spent in the southern part of the plain where the young animals feed on the rich growth of new grass. At these times the herds are widely spread over the grasslands, often scattered about in small troops.

FEW OR MANY OFFSPRING?
At the start of the dry season the small troops of grazers will move together, and start a long trek to the west. This is the rutting season – when the animals mate – and only the strongest males will have the energy to battle with their rivals, mate with the females, and still complete the journey.

July sees the animals on the move again, and the height of the dry season (August and September) is spent in the far north of their range, where they will remain until the first rains of late October signal the time to gather again for the trek south.

The peak of the calving season occurs in January as the migrating animals return to their southern feeding grounds. If the year has been exceptionally dry, the females will be unable to produce enough milk for their young, and few calves will be reared. Long dry spells also make the animals weak, so that many perish on the journey from exhaustion, disease and attacks by hyenas and hunting dogs. When rain is plentiful, however, the females are strong and thousands of calves are born and successfully raised. No matter how bad the previous years may have been, the population soon recovers.

POPULATION CONTROL
If an animal population becomes too large for the amount of food available, some kind of safety valve is needed to allow the species to survive. One simple means is for a proportion of the animals to die: usually the oldest, the very young, and any that are weak or sick. Alternatively the animals can move away, or migrate, in search of new feeding areas, as the Serengeti wildebeest do in Africa.

BIRD MIGRATIONS
Many birds in the far North breed during the brief Arctic summer, but fly south to more temperate areas in the winter. In the same way, many woodland birds of North America and Europe migrate to warmer regions during the winter months.

One of the spectacular sights for North American birdwatchers is the migration in the fall of birds of prey down the Appalachian Mountains on their way to their winter quarters in the southern States, Mexico, and Central and South America. Thousands of eagles, hawks and buzzards may be seen in a single day.

The journeys of many European songbirds are equally impressive. Before setting off on its annual migration to the savannah zone of Africa, a Sedge warbler may spend its last 3 weeks in northern France, fattening up on insects until it is almost twice its normal weight. Then, usually at night, it flies direct to southern Spain or North Africa, pauses there to feed and rest, and then makes a 30- to 40-hour non-stop flight across the Sahara Desert. At the finish of its 2,500mi journey, its stored energy reserves are completely used up. Huge numbers perish on the way from exhaustion, and in the talons of falcons that lie in wait for these migrants. Yet despite all the hazards, the warblers are back in northern Europe the next year. The numbers of individuals rise and fall, but the species survives.

FOOD SUPPLY AND NUMBERS
Other birds make less regular movements, but their numbers are still

▶ When the great wildebeest herds are on migration across the vast plains of the Serengeti they stop for nothing. Even when rivers are in flood the animals plunge in and swim across – though some of the younger and weaker ones may be swept to their deaths.

Control by disease
The European rabbit was brought to Australia in the 19th century, and within 40 years its population there was out of control. By the 1940s, loss of grazing meant that damage to the wool industry was costing $500 million a year, and lamb losses $200 million. In 1950, myxomatosis, a virulent disease found in South American forest rabbits, was introduced. The rabbit plague was wiped out.

controlled by food supply. Siskins, redpolls and crossbills all feed on tree seeds, but the amount of seed produced by most trees varies from year to year. As a result there may be four or five times as many of these birds one year as there were the year before. Instead of making well-organized migrations, the birds simply wander about until they find a suitable feeding area outside their normal range. Such movements are called irruptions.

Unlike trees, herbaceous plants produce roughly the same quantity of seed each year, and birds that rely on this source of food have much more stable populations. The number of linnets, for example, seldom varies more than 50 percent from one year to another.

HUNTERS AND HUNTED

Most predators have a preference for a particular type of prey. For example, Golden eagles in Europe feed mainly on rabbits and hares, while in North America they prey mainly on ground squirrels. Where mammals are hard to find, the eagle will hunt gamebirds such as grouse or ptarmigan. Each pair will claim a territory in wild open mountain or moorland country, and the amount of prey available will decide the size of that territory.

Where food is plentiful, a modest territory will provide all the food the birds need for themselves and their young. Where food is scarce a much larger territory will be needed. It may look as though the eagle population controls the number of rabbits or grouse, but the truth is more complicated – and it is usually the amount of food that determines the number of predators in an area.

▶ A locust swarm in Ethiopia. These insects are inconspicuous for much of their life, but when they swarm they take to the air in countless millions, often devastating entire food crops.

POPULATION EXPLOSIONS

Another factor controlling animal population size is the climate. Any gardener knows that the number of pest insects will be high after a mild winter and much lower after a long harsh winter. These changes, however, are small compared with the population explosions of the Desert locust of Africa.

In normal conditions, sparse dry vegetation is scattered over the floor of a desert valley. Here and there, well-camouflaged grasshoppers feed on blades of grass. Occasionally a pair will mate. A female lays her eggs beneath a bush, while some distance away young grasshopper larvae are struggling from the ground to disappear among the grass stems. But then comes a dramatic change. Clouds roll across the sky, and rain begins to fall, the first for over a year.

Within a week the valley is carpeted with vegetation. The grasshoppers mate more often, and their numbers multiply. Instead of avoiding each other, they form large groups. The young look different. Gone are the drab camouflage colors: all the new generations have bright black, orange and yellow patterns. A month later, the desert floor is covered in marching young grasshoppers. They molt, for the fifth and final time, and their transformation into winged locusts is complete. By now the whole valley is stripped of food, and the swarm takes to the air – up to 50,000 million of them rising together in a cloud so thick it almost blocks out the Sun.

Flying with the wind, the swarm drifts across country until it sees greenery. It then descends to feed – and an entire orchard or field of millet can be stripped bare in an hour.

The end may be just as sudden as the beginning. The wind may blow the swarm out to sea, or into a dry, barren area where there is no food; the insects perish in their millions. Back in the original valley, nothing stirs but a few drab, ordinary-looking grasshoppers waiting for the next rain.

The hare and the lynx

When one animal feeds almost entirely on one other, the populations of the two species might be expected to show a close relationship. And so they do. The population of the Snowshoe hare of North America has a natural ten-year cycle, possibly linked to variations in its own food supply, and this cycle is matched by a similar rise and fall in the number of lynxes – the hare's main predator. When the lynx population is small, Snowshoe hare numbers increase rapidly. Then, in response to the plentiful supply of food, the lynx population also increases. Eventually the number of hares goes down, the lynx population begins to decline, and the whole cycle is repeated.

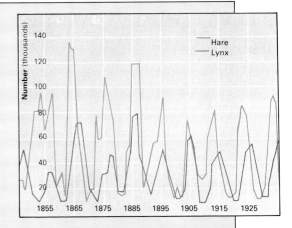

LIVING PARTNERSHIPS

In the heart of the Colombian rain forest a tiny White-tipped sicklebill perches on a bright red heliconia blossom. The bird is a hummingbird, and in order to feed it pushes its bill deep into the flower to reach the sweet nectar inside. As it does so its forehead becomes dusted with bright yellow pollen. The nectar is not a free gift. It is there for a special reason – and the pollen on the hummingbird's forehead is a clue. The bird is the heliconia's main pollinator.

The relationships between the hummingbirds and flowers of the South American rain forest are among the more complex partnerships that exist in the natural world. Others are rather more simple.

MARRIAGES OF CONVENIENCE
Some of the simplest associations in nature are those between climbing plants and the trees and other large plants they live on.

One characteristic of a climbing plant is that its main stem is usually slender and flexible. It is not designed to take the weight of the plant and so it does not need to be stiff and woody.

CLIMBING EQUIPMENT
Different climbing plants also have a variety of adaptations to help them hold on to their chosen support. Some, such as brambles, have hooks. Sweet peas and passion-flower vines have twining tendrils. Ivy has a mass of small air roots along its stem which can attach themselves to tree bark, rocks or stone walls. The Virginia creeper has small tendrils with flat adhesive stickers on the end. Climbing aids and long, slender stems are aimed at a very specific ecological niche. These plants put all their effort into climbing as high as possible to get their leaves into the sunlight and their flowers into the path of passing insects. Because they do not need great strength they can put their growing effort into length instead of strength – and passion-flower vines are known to reach lengths of up to 600ft.

HITCHING A LIFT
The climbing plants must not be confused with parasites. They make all their own food, take nothing from the host plant, and simply use the host as a means of hitching a ride upwards.

Very similar to the climbers are the epiphytes of the tropical forests, but instead of being rooted in the ground and climbing up their support trees, these plants never touch the ground at all. Their seeds are deposited on high branches, usually in the droppings of passing birds, and the plants often take root in the damp debris lying on the branch or in the fork of a tree. Other species simply dangle their roots in the moist forest air to obtain all the nourishment they need.

PLANT EATS PLANT
Plant parasites are a very different matter. A few are able to manufacture some of their own food, but most of them take all their food requirements from the host – and some even kill the host in the process. The mistletoes that grow on oak, apple and hawthorn trees are able to photosynthesize, but they also send a mass of roots into the branch of the host tree to "steal" additional nutrients.

Many fungi are totally parasitic on living plants, and their fruiting bodies – the reproductive parts – will often appear as large shelf-like plates or

---

Ant-plant partnerships
For millions of years, plants have been at war with insects and larger animals that feed on them. Some defend themselves with sharp thorns, others have poisonous stings or toxic leaves. Alas for the plants, animals evolve too. Goats and other browsers can eat the prickliest vegetation, while many caterpillars are immune from plant poisons – and even store them as a defense against predators. One intriguing defense is for a plant to provide a safe home for a colony of aggressive ants. The ants drive away plant-eating insects, and some species even cut away vegetation that gets in "their" plant's way. In return, the plant provides nest-sites and even food for the colony.

▶ *Pseudomyrmex* ants live inside the thorns of the Central American Bull's-horn acacia, and feed on the nectaries and modified leaf tips called Beltian bodies.

◀ The Southeast Asian epiphyte *Myrmecodia* houses colonies of ants inside its galleried potato-like tubers.

▶ *Crematogaster* ants in Kenya find a home inside the specially adapted thorns of some acacias.

▲ These huge elk-horn ferns are among the spectacular epiphytes that festoon the trees in the tropical rain forests of Queensland, Australia.

mushrooms on the sides of trees and other plants. They are a common sight in woodlands, but the attractive fruiting part hides a grim story. Inside the tree, the wood has been invaded in all directions by millions of fine white threads – the main part of the fungus. Slowly the tree is being killed, and its nutrients used by the parasite.

TEAM EFFORT
The largest flower in the world belongs to another parasite – the rafflesia of Sumatra. It has a bud the size of a soccer ball, opens with a hiss like a snake's, is over 3ft in diameter and has a smell that can knock you over. Its scent is the odor of rotting flesh – and is designed to attract the carrion flies that are the plant's pollinators.

The rafflesia has so many special requirements that it is remarkable that it has survived at all. It is a parasite on the root of a particular forest vine, and it requires not one but two animal helpers in order to survive. First it needs the carrion fly to visit its flower. The fly specializes in seeking out the rare rafflesia blooms, and as it feeds it carries pollen from one flower to another. Without this help the rafflesia would not be pollinated and could not produce seeds. But how does a rafflesia seed get within range of the underground root of a host vine to start a new plant? That service is provided by one of the larger forest animals such as a tapir or Sumatran rhinoceros. The seed sticks to its hoof and is trodden deep into the ground by the animal's weight.

PERFECT PARTNERSHIPS
Pollination and seed dispersal are the reasons for many of the associations that exist between animals and plants.

The hummingbirds are a perfect example of how evolution has given several South American forest plants all the help they need. A close look at their flowers reveals an amazing variety of shapes – and these shapes are matched by the bills of the birds that feed on them and provide the vital pollination service.

The sicklebill has a deeply down-curved bill to fit the shape of the heliconia flower. But the flower of the passion-plant *Passiflora mixta* is a straight, 5in-long trumpet with the nectar right at the end. The sicklebill could never reach it. Instead, this is the food-flower of the swordbill – a remarkable little bird with a bill 4in long – as long as its head and body combined! The two could not have started out with such extreme shapes, but must have evolved side-by-side over a very long time – the bill and the flower becoming gradually longer and longer in a process scientists call coevolution.

A PREDICTION CONFIRMED
A wonderful example of this effect is to be found on the island of Madagascar. There, during his travels, Charles Darwin (see page 9) found an orchid whose nectar was at the end of a tubular spur more than 6in long. The great man predicted that the flower could only be pollinated by an insect with a tongue that long. Scientists of the time laughed with scorn because no such creature was known.

Many years later the orchid's pollinator was discovered: it was a big Hummingbird hawk-moth with a 6-in tongue. The scientific world admitted its mistake by naming the moth *Xanthopan morgani praedicta*, in recognition of Darwin's prediction.

Many other animals act as pollen carriers. Fruit-eating bats pollinate certain tropical forest trees, and ants, beetles and wasps are helpers to other flowers. Even small mammals can help – like the tiny Honey possum of Australia whose furry nose gets covered in pollen as it feeds on the huge flowers of banksia plants.

TEMPTING COLORS
When it comes to spreading seeds around, animals are once again keen helpers – especially mammals and birds. Many plants enclose their seeds in brightly colored juicy berries and fruits which are very attractive to animals. These are eaten by birds and mammals, but the hard seed inside the fruit passes through the animal

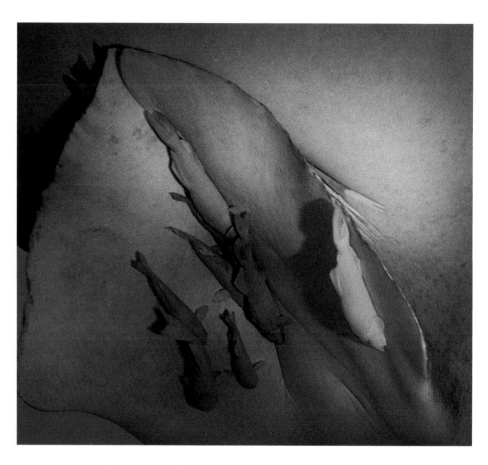

▲ Although the remora is quite capable of catching its own prey, it spends much of its time attached to a larger animal such as a shark, turtle or ray, feeding on the bigger animal's left-overs. The remora is attached by its dorsal fin, which is modified to form a sucker.

unharmed, and is deposited somewhere else in the animal's droppings.

Some seeds are so specialized that they actually need to pass through an animal in order to germinate properly. Seeds of the African beke tree, for example, will only germinate after passing through the gut of an elephant. They have an ideal start in life: landing on the ground already enclosed in warm, moist manure.

ANIMAL PARTNERS
The animal world has a similar range of partnerships, some of which, like those of climbing plants, are just

▲ No animal is too small to escape from parasites. Here, a yellow dung fly is infested with mites, which feed on body fluids after piercing areas of thin cuticle between the fly's body segments.

▼ The Common dormouse inhabits deciduous woodlands of Europe. It is an important disperser of plant seeds – especially those of hazel and beech which are among its favorite foods.

PARASITES: NICE OR NASTY?

There is probably not an animal in the world that does not have some other animals living on it or in it. Some are harmless, or just a mild nuisance. Some are very necessary – such as the bacteria that live in the mouth and gut and help to process food. But others are decidedly unwelcome. They range from the microscopic organisms that cause killer diseases such as malaria, cholera and typhoid, to internal parasites like tapeworms and liver flukes, and surface-dwelling animals like ticks, lice and fleas.

Even at this scale, some of these creatures are miracles of specialization. There are tiny mites that live inside the central shafts of birds' feathers, and miniature parasites that live on other parasites. There is a great deal of truth in the old rhyme (from the works of author Jonathan Swift):

"Big fleas have little fleas upon their backs to bite 'em...
Little fleas have lesser fleas, and so *ad infinitum*."

arrangements of convenience. The Hermit crab, for example, often has small anemones and other creatures attached to its shell. They probably help to hide the crab by camouflaging it, and in return they may benefit from food particles stirred up from the sea bed as the crab moves about. However, neither animal actually depends on the other.

Some associations can look quite dangerous. Certain small fishes live perfectly safely among the stinging tentacles of anemones and jellyfish. Apparently they are immune to the stings. Other small fishes called cleaner fish swim close to, or even inside, the mouths of large predatory or scavenging fish such as sharks and groupers. They are never harmed, and in return for food scraps which they pick up, they also help to clean irritating parasites from the larger fishes' mouth and gill areas.

PLANTS AND SUCCESSION

In 1883, Krakatau, a volcanic island between Java and Sumatra, was ripped apart by violent eruptions that were heard over 3,000mi away in Australia. Most of the island was destroyed, and the fragments that were left were stripped bare of every living thing. Fourteen years later those same rocks were clothed in vegetation. More than 60 different plant species had colonized the island, making a home for more than 130 species of insects and birds.

The story of Krakatau shows just how quickly a completely barren landscape can be colonized by plant and animal life. Only a few months after the huge explosions that destroyed most of the island, visiting scientists saw a tiny spider drifting by on its silk-thread parachute. The spider was almost certainly doomed, as the island was still a hot, barren wilderness of rock and volcanic ash, but plant seeds too were drifting in on the wind. Very soon, some of the toughest and most adaptable species would land, then germinate and stake their claim.

THE PIONEERS

These first arrivals are a very important group of plants. They are called "pioneer" species and they have a number of characteristics that enable them to survive where most other plants would perish.

Pioneer species are very hardy and can tolerate many extremes of soil and climate. Some are able to withstand drought and fierce heat and so are the colonists of the desert edges. Others can cope with extreme cold and windy conditions and so are able to colonize the higher reaches of mountains. Some are particularly tolerant of salt, and are the first plants to appear when a stretch of coastline becomes available for occupation.

New habitats do become available all the time. Volcanic eruptions create areas of dust, ash and solidifying lava. The end of an ice age or the retreat of a mountain glacier will expose bare ground that has been permanently frozen, perhaps for thousands of years. And down at the seashore, the shifting of a current or a river channel may uncover new sandbars or mudflats ripe for settlement by plants.

TRAVELERS BY AIR AND SEA

While many of the larger plants, especially trees, produce large, heavy seeds, the pioneers are mainly grasses and herbs which produce millions upon millions of seeds, mostly very small and light and with vanes, threads or parachutes to help them ride the wind. Research rockets sent up to sample the higher levels of the atmosphere have found that even many thousands of feet above the Earth's surface the air contains a surprising amount of plant and animal material. Rather like the floating plankton in the sea, though not nearly so rich, this aerial plankton contains seeds, plant debris, pollen of every imaginable species, and numerous flies, spiders and other tiny animals.

With so much living material drifting about on the global winds, it becomes easier to see how even the most remote island ends up teeming with animal and plant life. However, not all the colonists arrive by air: some are sea-borne. One famous marine traveler is the coconut. Although this nut is large and heavy it is also very buoyant and unusually resistant to long periods spent in salt water. It can drift for weeks on ocean currents, and throughout the tropics it is one of the commonest trees of coasts, islands and atolls.

The circulation of ocean currents also explains how many other animals find their way to remote islands. Sea turtles, and even some of the sea snakes, are able to cover enormous distances by swimming, but smaller animals such as lizards, frogs and mice occasionally arrive on rafts of vegetation that have drifted away from some distant swampy shore or have been swept out to sea on river currents.

◀This old ox-bow lake in the upper Amazon of Peru was formed when a meander (a tight bend in the river) became cut off. It is heavily colonized by floating and rooted water plants, and is slowly filling up with silt. Eventually, as the water becomes more and more shallow, trees will invade and the forest will cover the former pool completely.

THE PIONEERS GO IT ALONE

One of the most important characteristics of pioneer plants is to be found in the way they organize the production of new generations of their own kind. In some plant species there are completely separate male and female plants, while in others there are male and female parts on each individual plant. The pioneers all belong to this last group. There would be little point in taking root on a vast expanse of bare ground, only to die without producing any new plants to continue the line. By carrying with it all the nutrients and reproductive structures necessary to produce a new generation, just one seed can establish a plant species in a new habitat.

▼When glaciers retreat, they leave behind some very barren and unpromising landscapes, such as these bare rock faces on the edge of the Arctic. Lichens colonize the rocks, and slowly their dead remains collect in hollows and crevices where they form a thin soil in which patches of moss can grow. It does not amount to much – but it starts a new vegetation succession.

▲On the sandy shore of a tropical island in the South China Sea, Beach morning glory is growing in a continuous belt and colonizing the beach. Its seeds float on water and tolerate the sea's salt.

FIRST ON THE SCENE

One of the most familiar pioneer plants in the temperate regions of Europe and North America is the Rosebay willowherb, also known as the fireweed because of its preference for ground that has been burnt. This tall purple flower produces huge numbers of feathery-plumed seeds which are scattered by the wind, and it is often the first plant to colonize waste ground and the scorched ground exposed after forest fires.

SUCCESSIVE INVADERS

During their early stages, ecosystems are constantly changing. First the pioneers take root. They help to bind loose soil and stabilize it against erosion by wind and rain. They also add nutrients to the soil when they die, and humus (plant fiber), which improves the soil and helps to hold moisture in the surface layers where other plants can easily reach it.

Before long other plants arrive and become established. Many will grow faster or taller than the pioneers, and eventually they may take over completely. Over a long period of time, several quite different collections of plants may dominate the same area of ground one after the other, each one changing the soil, the amount of shade and the range of nutrients in the ground. This process of change is called a "primary succession," and it continues until a final steady state called a "climax" is reached.

A similar sequence of events takes place when an existing habitat is suddenly altered, either by a natural event such as a flood or a volcanic eruption, or by interference by man, such as the cutting down of a forest or the draining of a marsh. In this case the sequence of events is called a "secondary succession." The climax vegetation that results may be the same as the original climax, but the speed of the succession, and even the intermediate stages, may be different.

AN ECOSYSTEM EVOLVES

One of the main characteristics of any succession is that as time goes by the number of different plants in the community increases, and the total biomass – that is, the total amount of living material – also increases.

A small tarn in a highland area may start off as a still, clear pool surrounded by open ground covered with short grasses. Some time later the pool will contain numerous different water plants, with dense banks of reeds around the edges. Shrubs and bushes will then appear on the banks, and farther back there may be a scattering of small trees. Slowly, mud and decaying vegetation will fill the original tarn, and a few thousand years later, all that may remain of it might be a small clearing, carpeted in flowers and shrubs, and completely surrounded by dense mature forest.

SUCCESSION IN SAND DUNES

One of the best examples of a primary succession is the one that occurs on coastal sand dunes. Sand from lower down the beach is dried by the wind and Sun when the tide is out, and some of it is blown up the beach, beyond the high tide line, where it piles up into dunes. At first sight these mounds of sand look most unpromising as a site for colonizers. The coarse sand is forever on the move, threatening to bury any plant that does attempt to take root. It contains hardly any nutrients, and because there is no humus to act as a sponge, rainwater drains away through it almost as soon as it falls. This, combined with occasional flooding by storm waves, and

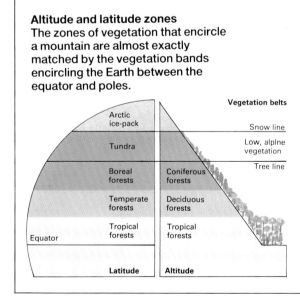

Altitude and latitude zones
The zones of vegetation that encircle a mountain are almost exactly matched by the vegetation bands encircling the Earth between the equator and poles.

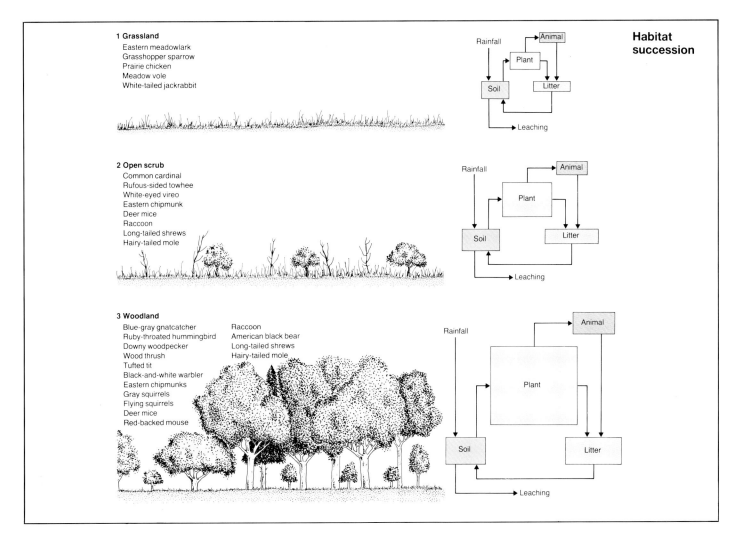

▲ The diagram above shows the increasing complexity and richness of a habitat succession from grassland to woodland in North America. As species diversify and biomass increases, the nutrient reservoir shifts from the soil to the vegetation.

constant exposure to wind and Sun, make the dune a harsh environment indeed. It is also very salty, something few plants are able to tolerate.

One species that is able to cope is Sand couch grass. It grows about 24in tall and spreads rapidly by sending out long underground stems, which help to bind the sand and stabilize the dune. The sand builds up around the grass stems, and once it is above the reach of the highest waves, other plants can start to move in. First to arrive is often Marram grass, a strong, fast-growing species which soon covers the surface. It grows to 4ft in height, and this has an important side effect. It causes the wind to swirl in eddies, dropping its load of blown sand amongst the grass so that the dunes quickly gain in height. Here we can see the ruthless side of succession in action, for once the Marram grass takes over, the Sand couch is doomed. Very soon it disappears completely, its job done. As the dune becomes more stable, new plants arrive. If the dune contains a high proportion of crushed seashell it will be rich in calcium, and shrubs such as sea buckthorn and privet will take root. If it is more acid, then pines may thrive.

ESSENTIAL ORGANISMS

One of the hidden factors in the development of an ecosystem is the importance of microbes. The upland tarn and the bare sand dune have virtually none of these essential organisms, so biological activity is very slow. Once the tarn has a variety of rooted weeds and tiny floating plants, and a layer of mud on the bottom full of decaying plant debris, the microbe population increases rapidly, and really gets to work. The same happens to a lesser degree in the sand dune once the surface is matted with roots and plant debris. The vital role of the microbes is to speed up the recycling process of the nutrients by breaking down plant and animal debris.

ZONES AND NICHES

Many ecosystems can be broken down into quite distinct subdivisions or zones, each with its own special group or "assemblage" of plants and animals in their select "niches" (see page 22).

Take, for example, a mountain near the equator, such as Mt Kenya, rising 17,150ft above the surrounding savannah. Its lower slopes are clothed in rain forest, ending at some 8,000ft. Above this there is a broad band of bamboo forest, giving way at 10,000ft to sub-alpine moorland covered with tussock grasses and lichen-covered tree heaths. Higher still lies a band of strange Afro-alpine vegetation – a dense carpet of mosses, lichens and short alpine grasses, dotted with outsized lobelias, and giant groundsels that grow up to 30ft high.

Each of these zones has its resident animals: buffalos, bongos and bush babies in the rain forest; monkeys, bushbucks and forest hogs in the bamboo zone; the occasional leopard prowling the sub-alpine zone in search of duikers and hyraxes; and above that, very few animals but for a few hyraxes and other rodents, ever on the alert for the hawks and eagles that patrol the bare mountain tops.

EARTH ZONES IN MINIATURE

The zones that typify the life of a mountain are almost exactly the same as the great bands of latitude that encircle the Earth parallel to the equator. The shapes of the continents and the global weather patterns cause many variations in the pattern, but broadly speaking the vegetation is the same. A traveler could begin his journey in tropical rain forest on the equator, move north through temperate deciduous forest before coming to coniferous forest. He could then cross the arctic tundra (very similar to high alpine vegetation) and finally arrive at the bare rock and ice of the polar zone. This habitat is very close to that of a high mountain peak.

ZONE SPECIALISTS

The fact that habitats are zoned means that the food available in each zone will be slightly different, and so will the physical conditions there. The animals that inhabit the zones will have to be suitably equipped, according to where they live. The Mountain ibex and chamois of Europe need thick, dense coats to keep them warm, and great agility to maintain their foothold on steep rocky slopes.

Life on a rocky shore is also clearly zoned, each zone having its own characteristic seaweed or combination of seaweeds. Animals here are also highly specialized. Limpets and barnacles must cope with spending half their time underwater and half exposed to wind and Sun. To avoid drying out and dying when the tide is out, many of these animals trap water inside their shells, and remain firmly shut until the sea washes over them again. Crabs and sand-hoppers avoid drying out or being eaten by seabirds by remaining out of sight under clumps of seaweed.

◀Hyraxes are small grazing animals whose "niche" is the rock outcrops (*kopjes*) that are dotted across the African savannah. Meagre grasses, seeds and lichen provide most of the animal's food, while rock crevices provide shelter from the midday Sun.

▼The saline marshes of the Camargue, at the mouth of the Rhône, are one of the last great wilderness areas in Europe. The marsh pools support numerous wading birds, each feeding in its own clearly defined zone and by its own specialized technique. Flamingoes filter food in the deeper water, dabbling ducks up-end in shallower areas, long-legged avocets and Black-winged stilts feed by sifting and probing respectively, while short-legged plovers scurry around at the water's edge.

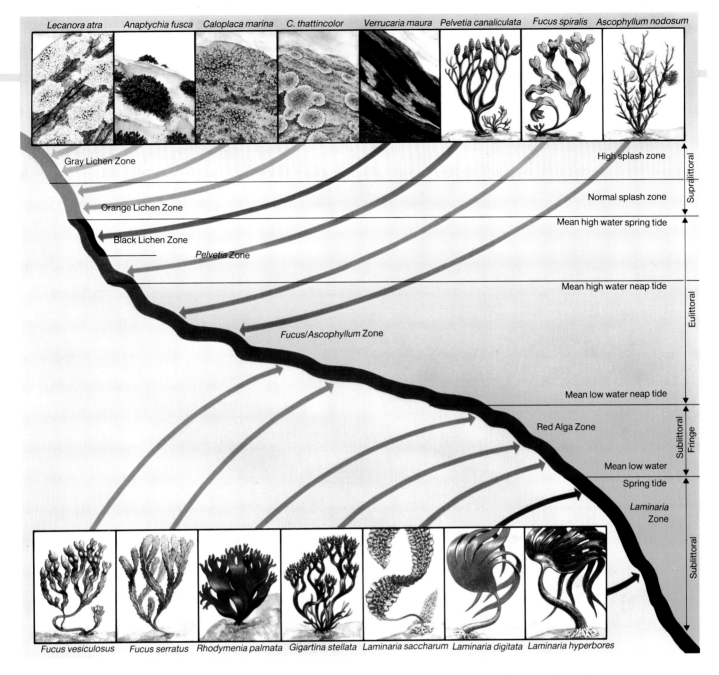

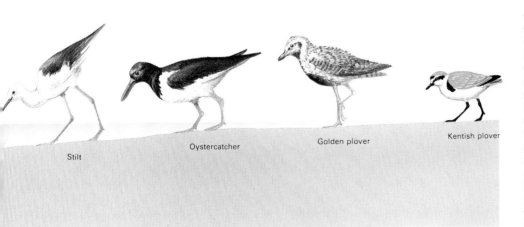

▲ Seaweed zonation

The most dramatic examples of zoning of plants are often found where two very different habitats meet, and no contrast could be greater than the meeting of land and sea on a rocky shore. The diagram shows the zonation and characteristic plants on a temperate rocky shore. Rocks at the top of the beach, beyond the normal tide line, are encrusted with lichens. Lower down are the familiar green and brown seaweeds such as bladder wrack, serrated wrack and spiral wrack all of which can tolerate quite long spells out of the water. The lowest parts of the shore are dominated by the red seaweeds and the huge kelp fronds which flourish only where they are permanently in water.

THE THREAT OF EXTINCTION

Somewhere in North America in about 1914 a hunter fired his gun, and for the very last time a Passenger pigeon fell to the ground at the hand of a human being. In the 1880s the skies of North America were full of these birds. Just 30 years later they had been completely wiped out.

A NATURAL PROCESS

The animals and plants that inhabit the Earth today may represent no more than one percent of those that have lived on Earth since life began over 3,500 million years ago. The fossil record shows that the average length of time on Earth for a mammal species is about 600,000 years. Some have lasted for much longer, but others stayed around for only a brief period of time before disappearing forever. Extinction, the death of a species, is itself a perfectly natural event. It is part of the slow process of evolution, and over a long enough period of time, every animal and plant species is likely to become extinct – either to have its line die out completely or to be replaced by a similar species that is better equipped to cope with the changing environment.

▶▼ Insects, the most numerous and diverse of all creatures (making up some three-quarters of the invertebrates), are also the most prone to extinction.
 Among the plants, many thousands of species are likely to be lost through the clearance of tropical forests. Here humankind stands to lose a great deal, for the tropical forests are home to many plants that are sources of life-saving medicines. We now know that for each mammal that becomes extinct, elsewhere on the planet we lose 2 birds, 4 to 6 fish, 70 plants and 180 insects.

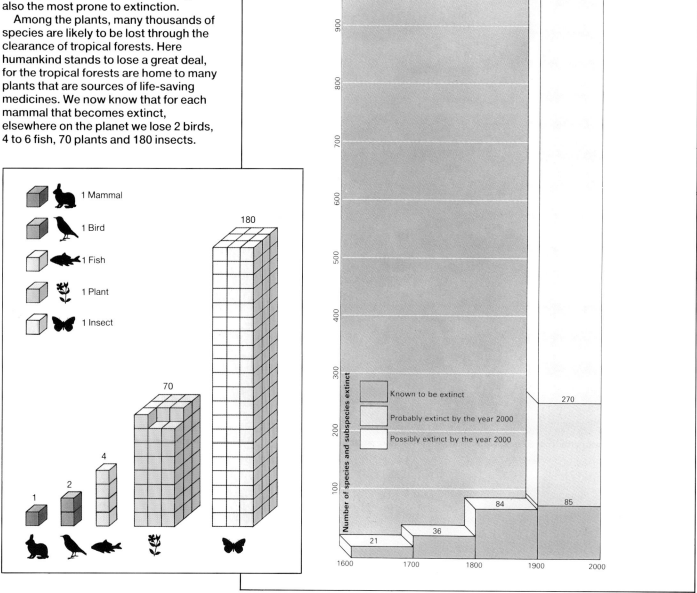

▲With modern earth-moving equipment, people can transform entire habitats. A huge dam project like this can devastate many square miles of land, and the lake that will form behind it may flood thousands of square miles and become a breeding ground for disease.

▼This tray from an insect collector's specimen cabinet gives just a hint at the number and variety of creatures in this huge animal group. Some may be carriers of disease, but many more are useful to us as they are predators on other insects that feed on crop plants.

DEATH OF THE DINOSAURS?

In the natural course of events, plants and animals become extinct for a number of reasons.

Some animal species may find themselves in competition for food or living space with another species that is stronger, bigger, or better adapted to the local conditions. They may be wiped out by the arrival on the scene of a new predator against which they have no defense or hiding place, or they may perish through disease.

In the past, many species became too specialized to cope with changes in climate, or the changes in sea level that occur when ice ages come and go. One fascinating theory suggests that the dinosaurs may have died out because of a change in the Earth's temperature caused by a huge meteorite hitting the planet and blanketing the atmosphere in dust and gas.

THE METEORITE MISHAP

The theory goes like this. In the majority of animal species that reproduce sexually, the sex of the offspring is decided at the moment the egg is fertilized. However, in alligators and crocodiles, the sex of the young is decided by the temperature of the eggs while they are being incubated. The female crocodile places her eggs in a nest-mound of damp grass and water-weed, which gets warm as the vegetation decomposes. The female croc will pile on more vegetation if she needs to increase the temperature, and scrape some away if the eggs start to get too warm. If she keeps the eggs at just the right temperature, roughly equal numbers of male and female crocodiles will hatch. If the temperature is a little too high, all the eggs will hatch out males: if it is a little too low, they will all produce females.

Now, the dinosaurs were reptiles too, and if their breeding biology was similar to that of these modern reptiles, it is just possible that their reproductive system could have been upset by a change in the Earth's temperature. If, over a period of time, nearly all the dinosaurs hatching out were of the same sex, they could have found it impossible to make up enough male-female pairs to keep the population going. In that situation their numbers would have dwindled quite rapidly to the point of no return.

THE PRESSURE OF MANKIND

We talk of the "mass extinctions" that wiped out the dinosaurs in just a few million years, and in geological terms that is a very fast rate of extinction, especially as the great reptiles had ruled the Earth for something like 140 million years before they suddenly vanished. However, that rate is slow compared with what is happening on our planet today.

Current rates of extinction, and the numbers of animals now under threat, are far higher than at any time in our planet's long history, and these events are not natural: they are caused mainly by the activities and increase of just one species – our own.

The human species is far more intelligent than any other that has ever existed. We have huge powers of learning, and of communication. We can invent things; we build cities, roads and rail systems; we organize ourselves into complex societies, and we make long term plans. Perhaps most important of all, our numbers have grown faster than those of any other animal – and they continue to grow at an alarming rate.

The result is that now enormous demands are made on the natural resources of the Earth. Just a few million years ago, humans were weak, timid creatures gathering fruit, berries and roots or hunting small animals. Now we completely dominate the world, and this of course affects the plants and other animals that share our planet.

UNKNOWN LOSS OF SPECIES

In the last 400 years, something like 34 mammal species and 94 bird species have become extinct, and today the International Union for the Conservation of Nature (IUCN) lists at least 120 more mammals and nearly 1,000 birds as being under serious threat of extinction (see also page 46).

One of the greatest tragedies of all is that we still know only a small proportion of the plant and animal species that live on Earth. About 1.6 million

▲▶ The thrush family contains a great variety of insect-eating birds. Without these busy predators the world's insect population would soon be out of control. This illustration shows a Gray-headed parrotbill (*Paradoxornis gularis*) (1), a Yellow-rumped thornbill (*Acanthiza chrysorrhoa*) (2), a Gray fantail (*Rhipidura fuliginosa*) (3), an African paradise flycatcher (*Terpsiphone viridis*) (4), a Black-backed fairy-wren (*Malurus melanotus*) (5), a Golden whistler (*Pachycephala pectoralis*) (6) and a Northern logrunner (*Orthonyx spaldingi*) (7).

species have been named so far, roughly half of them insects and about one-fifth plants, but every year at least 5,000 new species are discovered. Estimates of the total number of species yet to be discovered vary between five and ten million, but many of these are doomed to disappear before we even know them. Some scientists predict that we may lose a million species during the last quarter of this century, and if that prediction is correct, it means the loss of 100 species every day – or roughly one every 15 minutes.

THE THREAT OF THE HUNTER
Mankind poses many different threats to the natural world. Some are caused by greed, lack of thought or ignorance and others by situations that may be beyond our control.

In the age of sailing ships, many remote islands provided safe anchorages where ships could stop to stock up on fresh water and food. Few of the islands had any large predatory animals because it is difficult for them to cross the open seas, and for this reason many of the islands were home to large flightless birds. These were quite tame and provided an easy source of food for the hungry seafarers. The moa of New Zealand and the dodo of Mauritius were just two of the many species that were wiped out in this way before anyone realized they were even in danger.

Other species have been regarded as a threat to people, or to their crops or farm animals, and for that reason they have been persecuted.

The Gray wolf was once widespread throughout the Northern hemisphere, but it was hunted to extinction in Britain and many parts of Western Europe and is now rare. It survives only in areas of northern North America, northern Asia and in those parts of the Middle East where there are relatively few people to harass it. In Europe and North America

41

the Golden eagle was branded a sheep-killer in the last century, and for many years it was shot on sight, or killed by farmers putting out poisoned bait. In the United States the slaughter continued right up until 1962. Then, a new law put an end to the shooting, which for over 20 years had claimed up to 2,000 birds every year in the sheep-ranching areas of Texas and New Mexico.

The same treatment was handed out to the White-tailed eagle in Scandinavia and the Wedge-tailed eagle in Australia. In most of these cases the birds were largely innocent. The Golden eagle probably does occasionally take a lamb or a weak or injured sheep, but its main prey consists of rabbits and hares, and ground-nesting game birds.

▼ By a chance alteration of the gene code, the African white rhino has two horns while the Indian rhino has just one. Sadly, both species are hunted, illegally, because of the supposed magic properties of rhinoceros horn.

THE DEADLY TRADE

One of the least forgivable threats to animal life is the trade in live specimens of rare animals and in their skins, feathers, horns and shells. In the 19th century the beautiful Chinese egret was almost wiped out because of the fashion for hats and cloaks to be adorned with their elegant plumes. Many other birds, including some of the spectacular birds of paradise, also came close to extinction.

This deadly trade eventually caused such an outcry that in 1922 the International Council for Bird Preservation (ICBP) was born – the world's first major international conservation organization. Sadly, however, this illegal trade in live birds continues. Collectors will pay up to US$20,000 for a single specimen of a rare parrot

▶ Whaling has been a traditional way of life for centuries in the Faeroe Islands, halfway between Britain and Iceland. Here, pilot whale carcasses are lined up on the beach awaiting transfer to the processing factory.

or macaw, which is why so many of these Latin American forest birds are on the IUCN "Red List" – the catalog of the world's threatened species.

Mammals and reptiles also suffer from widespread illegal hunting and trading. Elephants of all sizes are killed for their ivory; Clouded and Snow leopards, ocelots and other spotted cats for their furs; and crocodiles, alligators and snakes for their skins used to make shoes and bags.

In many of the poorer countries of the world it is a common sight to find skins, shells and stuffed animals for sale to tourists from roadside stalls. Here too, though, the tide is turning. Public opinion in much of Europe and North America is now against the wearing of skins and furs and the buying of "trophies" and souvenirs such as seashells.

NO EXCUSE WILL DO

Many people argue that they are doing no damage because the animal they buy is already dead. That excuse simply will not do. Each time someone buys a stuffed animal, another is killed to take its place on the market stall. If people refused to buy these goods, the traders and the hunters who supply them would have no market. The deadly trade would itself soon face extinction. Yet despite the Convention on International Trade in Endangered Species of Wild Flora and Fauna (CITES), which has now been signed by more than 90 countries, illegal trading in rare and endangered animals is still a major threat. And the problem is not confined to just developing countries.

▼By opening its mouth, this Orinoco crocodile allows the Sun to warm the inside of its mouth where blood vessels lie close to the skin surface. Like most crocs it is hunted for its skin.

OUR PROGRESS – OR DESPAIR?

By far the most serious of the man-made threats to the plant and animal life of our planet is the wholesale destruction of habitats.

In the name of progress we may "reclaim" an area of marshland or cut down a forest. There is always a "good reason." We need the space for agriculture, for housing, or for a new airport. Unfortunately, we often overlook the long-term damage that such actions can cause.

HABITATS GONE FOREVER

The Waddenzee in the northern part of the Netherlands, is the largest complex of coastal habitat in northern Europe. Its varied ecological niches are essential to the plants, birds, insects and coastal sealife.

Yet in recent years large areas have been reclaimed. The offshore islands are rapidly becoming overloaded with tourists and the roads, hotels and other services they require, and the Waddenzee itself is being poisoned by pollutants swept into it from the industrial regions of Germany, France and the Netherlands by the Elbe, Ems, Rhine and other rivers.

Habitat destruction is happening throughout the developed world – and often in areas where already there is hardly any of the original natural habitat left.

It is quite wrong to assume that all rare, fascinating and potentially useful species live far away in tropical rain forests. Many of them do – but by no means all. For example, the El *segundo* butterfly occurs in only two places in the world, both in Los Angeles County, California. One is in the middle of a Standard Oil refinery, the other right next to Los Angeles airport!

THE THIRD WORLD TRAP

The problems of habitat destruction in the Third World are often much more difficult to tackle than those in the developed world. It is easy to criticize, but in many of the poorer countries of the world the people have no alternative. If it comes to a choice between cutting down forest to grow crops, or seeing his children starve, any peasant farmer in Asia or South America will cut down the trees in order to feed his family. The same is true in the dry lands of Africa.

LANDS MADE LIFELESS

Today we see countries such as Mali, Chad, Ethiopia and Sudan as drought-ridden dustbowls. It is hard to believe, but less than 100 years ago large parts of these countries were covered in open woodland and grassland.

The damage is largely due to the enormous growth of the human population and its demand for food,

▲ A lone Gray wolf pads along through the snow. Most wolves live in packs, so this loner is probably a young male in search of a territory and a mate: he will skirt the territories of other wolves.

fuelwood, and large areas of grazing land for cattle and goats.

Years of drought have tipped the situation into disaster, but if the original vegetation had been intact, the droughts would have been less severe and the land would have been able to recover. Stripped of its vegetation cover, the ground is parched by the Sun, and before long it is reduced to desert. The people lose what land they had, and along with its natural vegetation the region loses most of the birds, insects, mammals and reptiles that used to live there.

THE TIDE OF EXTINCTION

In the last few hundred years, as many species have been lost as in the whole of the last 2,000 years, and at present rates we could lose as many again in the next 50 years. The chart below shows a selection of extinct and endangered species. However, many sub-species are also in danger. For example, the world population of leopards is quite healthy, but at least five sub-species are now in danger.

Many birds are facing the same threat. Worldwide the kestrel has many healthy populations, but the Mauritius kestrel, found only on the island whose name it shares, is among the world's rarest birds of prey. In 1974 only six were known to be on the island, and even after concentrated conservation efforts the 1988 total was still only about 25 birds in the wild and handful in the island's captive breeding enclosures.

▼ One of the most striking features of the time chart below is the number of island species it contains. The reasons are varied: the impact of alien competitors and predators introduced by people, either deliberately or by accident, the loss of habitat due to agriculture and forestry operations, the effect of introduced diseases and the huge growth of international tourism.

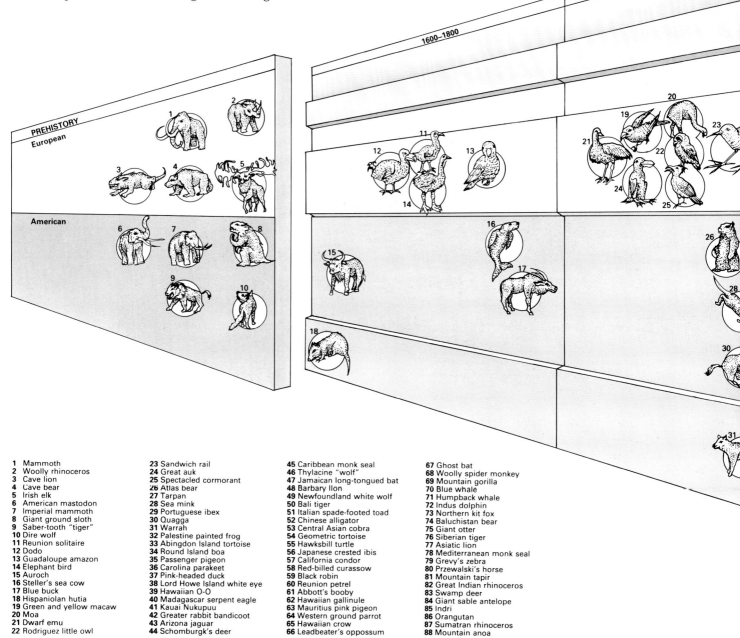

1 Mammoth	23 Sandwich rail	45 Caribbean monk seal	67 Ghost bat
2 Woolly rhinoceros	24 Great auk	46 Thylacine "wolf"	68 Woolly spider monkey
3 Cave lion	25 Spectacled cormorant	47 Jamaican long-tongued bat	69 Mountain gorilla
4 Cave bear	26 Atlas bear	48 Barbary lion	70 Blue whale
5 Irish elk	27 Tarpan	49 Newfoundland white wolf	71 Humpback whale
6 American mastodon	28 Sea mink	50 Bali tiger	72 Indus dolphin
7 Imperial mammoth	29 Portuguese ibex	51 Italian spade-footed toad	73 Northern kit fox
8 Giant ground sloth	30 Quagga	52 Chinese alligator	74 Baluchistan bear
9 Saber-tooth "tiger"	31 Warrah	53 Central Asian cobra	75 Giant otter
10 Dire wolf	32 Palestine painted frog	54 Geometric tortoise	76 Siberian tiger
11 Reunion solitaire	33 Abingdon Island tortoise	55 Hawksbill turtle	77 Asiatic lion
12 Dodo	34 Round Island boa	56 Japanese crested ibis	78 Mediterranean monk seal
13 Guadaloupe amazon	35 Passenger pigeon	57 California condor	79 Grevy's zebra
14 Elephant bird	36 Carolina parakeet	58 Red-billed curassow	80 Przewalski's horse
15 Auroch	37 Pink-headed duck	59 Black robin	81 Mountain tapir
16 Steller's sea cow	38 Lord Howe Island white eye	60 Reunion petrel	82 Great Indian rhinoceros
17 Blue buck	39 Hawaiian O-O	61 Abbott's booby	83 Swamp deer
18 Hispaniolan hutia	40 Madagascar serpent eagle	62 Hawaiian gallinule	84 Giant sable antelope
19 Green and yellow macaw	41 Kauai Nukupuu	63 Mauritius pink pigeon	85 Indri
20 Moa	42 Greater rabbit bandicoot	64 Western ground parrot	86 Orangutan
21 Dwarf emu	43 Arizona jaguar	65 Hawaiian crow	87 Sumatran rhinoceros
22 Rodriguez little owl	44 Schomburgk's deer	66 Leadbeater's opposum	88 Mountain anoa

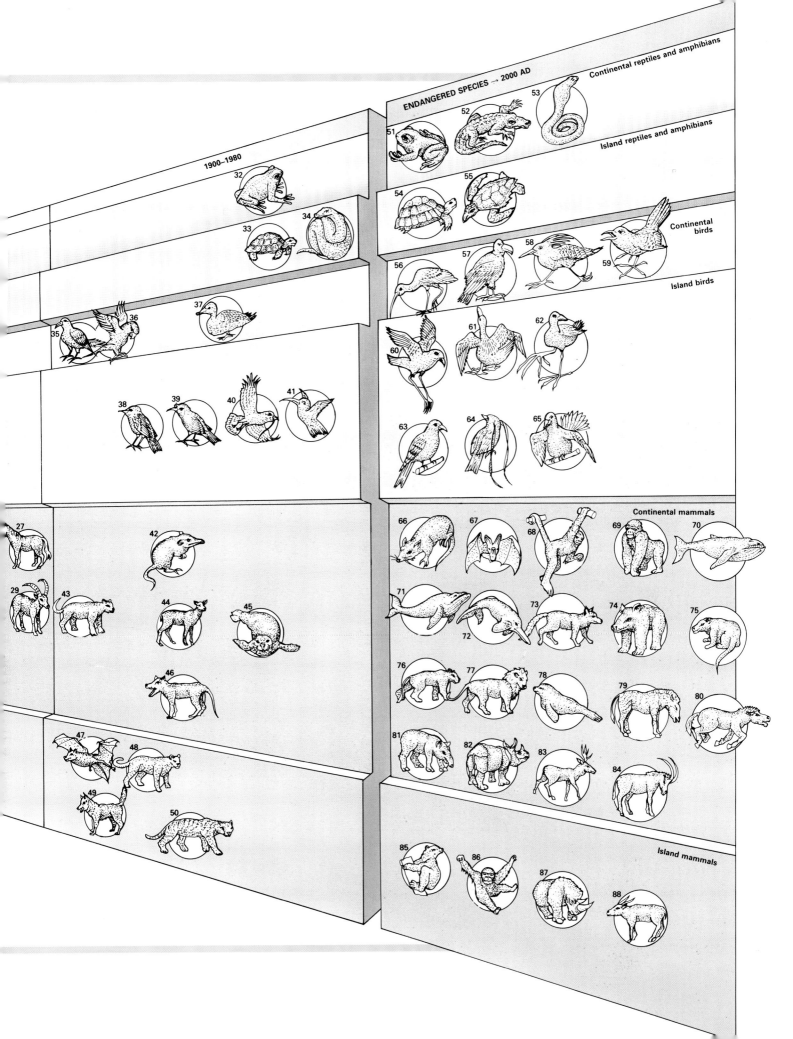

THE PRESSURE OF PEOPLE

With more than 80 million people, Nigeria now accounts for almost one-sixth of the total population of Africa. By the year 2010, the figure will have risen to at least one-fifth, and Africa as a whole will have to find ways of feeding an extra 694 million people. Human population growth lies at the heart of many world problems. Today we number almost 5,000 million. By the year 2000 we will have topped 6,000 million.

HUMAN OVERPOPULATION
There are two main reasons why the human population has grown at such a phenomenal rate.

Firstly, the development of agriculture and then of industry made it possible for large numbers of people to live together in communities, with different people doing different jobs, all contributing to a local economy that provided everything that each person needed.

The second factor is that as humans turned away from the hunter-gatherer lifestyle to a lifestyle based on living in towns and cities, their attitude to child-rearing changed dramatically.

THE BIRTH OF FARMING
Whoever first pushed a seed or a sprig of vegetation into the ground in the hope that it would grow into a new plant probably took one of the most significant steps in the history of humankind. It was the first step in the development of agriculture.

The first cultivators probably did no more than plant seeds, roots or tubers of some of their favorite food-plants.

▶ The graph of world population shows a rate of increase that will take the total past the 6,000 million mark by the year 2000. Inset diagrams illustrate the typical population profiles of developed and developing regions.

This would have given them a more reliable supply, nearer to home, and would have saved the time they would otherwise have spent searching for plants growing wild. Later, someone would have had the idea of planting only the seeds, roots or tubers of the biggest and strongest plants, and of those that produced most food, or the tastiest fruit. In its simplest form this was the start of selective breeding of crops.

WHY WERE FARMERS NEEDED?
We know now that large-scale cultivation started in the Middle East, about 10,000 years ago. It provided the people of that region with a much greater amount of food, and also ensured a steady supply, but there is an intriguing question over precisely why people switched from hunting and gathering to growing crops and tending herds of livestock. It used to be thought that farming was an easier way of life than hunting and gathering. Modern evidence shows that exactly the opposite is true.

The hunter-gatherers (Bushmen) of the Kalahari Desert in South West Africa feed their small bands quite easily by working about 6 hours per day, 3 days per week. Old people and children do no work at all. And yet these tribesmen live in one of the

▶ *Homo sapiens sapiens* – in terms of numbers of individuals, range and ability to survive in any environment "modern humans" are the most successful species ever to have evolved on Earth.

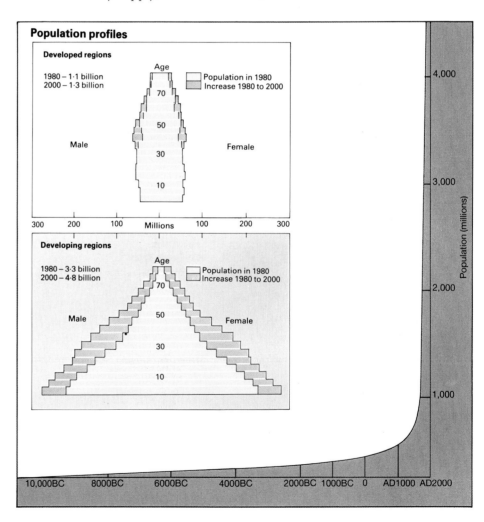

▲A shanty town in Ecuador. In South America, where the birth rate continues to soar, many major cities are unable to accommodate their swollen populations. For millions of people, home is a shack of cardboard or corrugated metal sheet, somewhere outside the city, often on the edge of a municipal refuse tip.

harshest environments on Earth. In the Middle East, where there was abundant fish, game, fruit and wild grain, the hunter-gatherer way of life should have been extremely easy. Instead of *choosing* agriculture, it seems much more likely that events *forced* people into farming.

Cultivation produces much more food than nature alone, and this plentiful food supply would have allowed the human population to grow very quickly. In a relatively short space of time it would have reached the point at which intensive cultivation *had* to continue, or people would have gone hungry. Cultivation was already demanding more labor – and that was to become a key factor in the human population boom.

INCREASING PRODUCTIVITY

The development of agricultural science and technology has greatly increased food production over the past 200 years, and especially over the past few decades. Primitive farmers may have produced at most about 2 tons of wheat for each acre of land. British farmers in 1900 probably obtained about 5 tons for each acre. Today, the highly mechanized farms of Europe and North America average 15 tons for each acre. Also, farmland has spread so much that arable land now occupies 10 percent of the total land surface while pasture occupies another 23 percent. Hand in hand with this growth in food productivity, the world population has increased threefold since 1850.

FARMING AND FAMILY SIZE

In hunter-gatherer societies there is a desire to limit the size of the family. People wish to have children, and there is usually plenty of food with which to feed them, but they have no wish – and no need – for very large families. Traditional farmers tend to have large families because working the land is so labor-intensive.

In the affluent and highly mechanized countries of Europe, Australasia and North America, many farmers can afford to employ labor. In the Third World very few farmers have that option. They simply do not have enough money to employ labor, and so the workforce must come from the family itself.

Another important factor in the Third World is that few countries have any form of welfare state. The only form of "insurance" people have against the problems of old age is their children, who will then take care of them. This is why, today, more than 80 percent of all world population growth is taking place in the poorer countries of the world – in Latin America, Africa and parts of Asia.

PATTERNS OF POPULATION
When people in Europe started to move into towns and cities during the industrial revolution of the 19th century, they initially kept to the old rural tradition of having large families. At that time urban housing and general living standards were very poor. Factories employed children as young as nine years old, and all but the very young and the very old had to work as every available cent of income was needed to help feed and clothe the family. However, as living conditions improved throughout the industrialized world, population figures changed dramatically.

The availability of better housing, a healthy diet and medical care meant that people began to live much longer. As the death rate began to fall, so did the birth rate. People no longer produced large families, and as the number of deaths was almost exactly balanced by the number of births, the population of many countries stabilized. Today the typical population profiles of developed and developing countries are very different. In developed countries the number of young people is roughly the same as the number of middle-aged people.

In developing countries the young far outnumber the middle-aged, and as these young people will before long produce children of their own, the result will be another sudden rise in the population figures.

CONTROLLING GROWTH
The population of South America is about 300 million people, and if it continues to increase at its present rate, in another 500 years the entire continent would be filled to capacity. However, long before that stage could be reached, millions of people would die from famine and disease. If this grim prediction is to be avoided, the people themselves must be persuaded to have fewer children.

In some countries, such as India and China, governments have already decided that population growth must be brought under control. (India has about 750 million people; China has around 1,000 million.) Many other countries have not yet made this important decision. In recent years,

▼ In many parts of the world, people consume excessive amounts of nutrients, which leads to a prevalence of obesity, dental caries and heart disease. The developing world's nutritional disorders – notably anemia and malnutrition – are primarily related to nutrient shortage. "Poor diet" in general usually refers to a low level of "dietary energy supply" (DES). According to FAO figures, average requirement is 2,092 kcals/day. Below 1,765 kcals/day, people are starving.

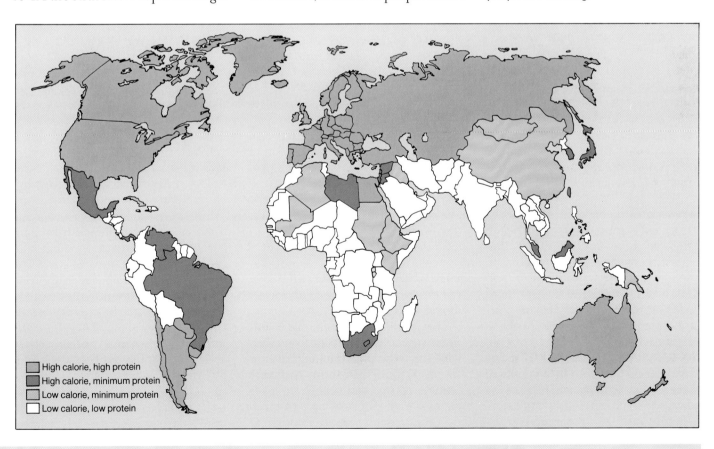

High calorie, high protein
High calorie, minimum protein
Low calorie, minimum protein
Low calorie, low protein

▲ !Kung hunter-gatherers of the Kalahari returning from a hunt (Southern Africa). These people are well aware of the techniques of cultivation, but find the hunter-gatherer life-style better suited to the resources of their environment.

for example, Tanzania has argued that her population is still too small, and that she needs more people to work the productive land that is available. Argentina has claimed a need for more people in order to guarantee her national defense. In other countries, especially those in South America, there are strong religious objections to the idea of birth control.

THE NEED FOR BIRTH CONTROL

Birth control also presents many practical difficulties. The first is a problem of education. Introducing birth control often goes against very strong cultural traditions, and in areas where there are few schools, and most of the adult population has little education, the first task is to get across the message of why population control is necessary. This means training people to go out into the rural areas and teach the local people about health, hygiene, and improving their diet and so on, as well as about birth control.

Next there is the problem of how to put birth control into practice. Methods of contraception that are safe and reliable are also very expensive. They are simply not available in the remote villages and sprawling slums of many Third World countries – yet this is where the need is greatest.

Even in India, where the need for birth control has already been accepted, the population continues to rise rapidly and is expected to top 1,000 million by the year 2000.

Despite the efforts that are being made to tackle population problems, the number of people in the world will continue to rise. Even if there was worldwide agreement to limit families to no more than two children, and if contraception was freely available in every country, the total world population will still pass the 6,000 million mark by the end of this century.

HOW MANY CAN BE FED?

Whether or not there are too many people in the world, or in any particular country, is often the subject of bitter argument. Yet with the right organization it is theoretically possible to feed the entire world from the amount of productive land we have at present. In fact some scientists estimate that it would be possible to feed three times the present world population. The problem is not that there is no food, but that the food we can produce is not in the right place at the right time.

The fertile farmlands of North America, Europe, Australia, New Zealand and large parts of Asia produce far more food than those countries require. A large part of the surplus is exported, as a normal part of international trade. Some is shipped to parts of the world where people are hungry, as part of the developed world's aid program. But there are problems with agricultural overproduction as well.

Some goods have a very short life and cannot easily be transported to areas where they could be used to combat hunger. Others could be transported if refrigerated containers were available and if refrigerated storage was available in the country of destination. Sadly many countries simply do not have these necessary facilities. So while grain, sugar, dried milk and a number of other products can all be used in relief programs, most dairy products, fruit and vegetables would simply spoil long before they reached their destination.

The greatest irony of all is that for various commercial and political reasons, huge quantities of surplus food go to waste every year – dumped, buried, burned, or at best plowed back into the land they came from.

THE ULTIMATE GOAL

In the short term, international food-aid programs will continue to play a vital role in fighting world hunger. But the long-term goal must be to help the poorer nations feed themselves. Aid will still be very necessary, but it will have to focus more and more on providing educational facilities, technology, financial aid and production equipment.

Educational aid may be given in the form of money to be invested in schools, health education, adult education programs and in technical training. It may also include supplying teachers and administrators to help set up the systems the country needs.

▶ Much of the fish taken by British ships in the Arctic and North Atlantic fishing grounds never comes ashore in Britain. It is bought direct from the catchers as they lie at anchor off the Scottish coast, mainly by huge factory ships from the USSR and other east European nations.

Suitable technology will include help in finding and tapping new underground water resources so that irrigation can be used in regions that are dry but have potentially productive soil. It will also include supplying suitable fertilizers, and new strains of food-crop seeds that are better able to cope with drought, or with soil that is too salty or contains high concentrations of certain toxic minerals. Help will be needed to improve harvesting methods, and food storage methods.

WASTAGE AND DEBT

It is tragic that in the Third World, where people are already short of food, up to 10 percent of the grain crop is lost every year, after it has been harvested, from insects, molds, rats and other storage problems. The quantity is estimated at well over 70 million tons per year of essential food grains such as rice and maize. Even quite simple, low-tech improvements can often halve a farmer's losses.

Finally, financial aid will have to tackle the crippling problem of Third World debt. Some countries now owe so much that the repayment instalments alone are much more than the country earns or produces in a year. In that situation the country cannot hope to invest in better agriculture, education, food supply systems or roads. Sooner or later the large international banks and lending countries will have to write off these debts and adopt a "fresh start" policy.

DESTRUCTION OF THE LAND

Too many people putting too much strain on poor land is a recipe for disaster. In many parts of Africa the problem lies in the herds of goats and cattle that strip the land of even the thorniest scrub vegetation. In other areas it is the search for fuelwood that is the main cause of environmental damage. From the comfort of Western Europe or America it is hard to understand that almost half the world's population – more than 2,000 million people – depend directly on firewood to warm their homes and cook their food. In large areas of the world the land has been stripped bare of wood for great distances. The result is always the same. Once the soil is exposed in this way it is baked by the Sun, and erosion starts with the very first rains. In a very short time the

▲ The vast cereal surpluses of North America and Western Europe have often been both commercially and politically embarrassing. Some grain is used in aid programs, but much is wasted.

▼ By the year 2000, fuelwood supplies of an estimated 2,400 million people will either be below minimum requirements or so low that they can be maintained only by cutting wood faster than natural regeneration can replace it.

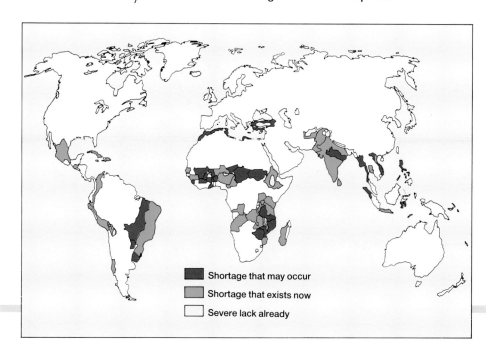

fragile soil is washed away, leaving a hard-baked surface on which nothing can grow. Meanwhile the precious topsoil is being swept along in muddy rivers and dumped into the sea. Eventually, the people are left with no choice but to move on – and repeat the damage.

BREAKING THE VICIOUS CIRCLE
Many countries, and major international organizations such as the UN Food and Agriculture Organization, have schemes to try and break the cycle of land degradation.

Building terraces and planting fast-growing trees can stabilize mountain slopes stripped of their vegetation.

In semi-desert regions, fast-growing species of *Casuarina* and *Acacia* can provide fuelwood for people and foliage for animal feed. The trees also protect the soil from erosion, enrich it by fixing atmospheric nitrogen, and provide cover for other crop plants.

The destruction of rain forest by migrant farmers can also be reduced by introducing new methods. Instead of clearing large areas of forest, for planting with manioc or maize, which exhausts the soil in only about 2 or 3 years, the new methods copy the natural forest structure by mixing rows of beans, groundnuts, sweet potatoes or pineapples with taller banana, coffee and oil palm plants.

In some areas fish farming in man-made ponds has been introduced into the "multiple use" forest schemes; in others, wild forest products such as Brazil nuts and honey are collected and sold. All these systems aim to replace destructive practices with new ones designed to provide long-term "sustainable" food production without damaging the environment.

▶A man and woman make their way back to their village in the Pokjhora region of Nepal with heavy loads of fuelwood. In many poorer countries the desperate search for fuel can take up most of the working day.

DESTRUCTION OF FORESTS

In the heart of the Malaysian rain forest a logger switches off his chainsaw as a mighty forest tree sways and then, with a crack like a rifle shot, begins to fall. The huge trunk hits the ground just 20 minutes after the logger started work. In that time, worldwide, an area of tropical forest the size of 620 full-size soccer pitches has been cut down.

▼Forest clear-cutting like this in the Amazon threatens countless plants and animals, and may also cause climatic changes by upsetting the water cycle.

Each year the total area of tropical forest lost on our planet is equal in size to the mountain kingdom of Nepal. If the present rate of cutting continues, the tropical forests could be lost forever in the next 50 years.

SLASH-AND-BURN FARMING

In many tropical forest regions, the ancient form of agriculture known as slash-and-burn accounts for huge amounts of forest loss. The migrant farmer cuts down a small patch of forest, burns the trees and then plants his crops. The ash from the fire acts as a fertilizer and enriches the thin forest soil, but even with this help the soil is exhausted after only 2 years and the farmer and his family are forced to move on.

In the past, the number of slash-and-burn farmers in the tropical forests was not too great. The cleared plots were like pin-pricks, and the forest had plenty of time to recover when the farmers moved on. Today, however, the migrant population has grown enormously and there are more slash-and-burn cultivators than the forest can accommodate. Large areas of forest are now patchworks of partly regrown abandoned plots, with little untouched forest in between.

THE IMPACT ON WILDLIFE

As virgin forest is cut down, the effects are felt through the entire forest food-web. Predators such as the Amazon's Harpy eagle and jaguar need large territories in which to hunt. If they are disturbed by intruders they often fail to breed successfully. Other species move constantly, feeding on the rich resource of fruiting trees; but logging operations break up the canopy into "islands" separated by areas of open ground which the animals either cannot or will not cross. This in turn breaks up their populations into small and isolated units that are vulnerable to fires, outbreaks of disease and inbreeding.

RUINOUS CATTLE RANCHING

The Amazon rain forest has suffered in particular from felling and clearing to make way for herds of cattle. Every year, many thousands of square miles of virgin forest are burned to provide poor-quality pasture, yet cattle-raising on the thin tropical forest soil is even more wasteful and destructive than slash-and-burn farming.

In the past the cattle-owners have been encouraged by government grants – and by a huge international market for the cheap beef they export. But the real price is high. The pasture is exhausted after only 6 years, the ground trampled, and burned by the Sun. The forest cannot grow again, and the former pasture is abandoned as a dry wasteland.

Disasters like this in the world's poorer countries are often fuelled by demands in the affluent countries of Europe and North America. But public opinion is a powerful force, and in the United States in recent years a lead was taken by the fast-food chain Burger King when it announced that it would no longer buy meat raised on rain forest land. Since then other food organizations have followed suit.

THE SPREAD OF DESERTS

The last two chapters showed how population growth is putting a huge strain on natural resources and habitats in many parts of the world. From many newspaper reports and television pictures it is easy to believe that vast areas of eastern Africa, and the Sahel region south of the Sahara, are natural desert lands (see page 45). They are not. It is the pressure of people, and their herds of cattle and goats, combined with a series of severe droughts, that has reduced them to parched deserts.

EROSION OF MOUNTAIN LAND

The effect of human pressure is also clear in the mountain countries of the Himalayas. Here the number of

people per square mile may be much lower than in parts of Africa, but their impact is just as severe.

As much as 60 percent of Nepal's population lives in the hills, and one of the main sources of food for the people here is the buffalo, whose milk is an important part of their diet. But a single buffalo requires 40lb of fodder a day. The hills are now so bare of leafy vegetation that gathering and carrying that amount can take a large part of the day. Stripped of their forest cover the hillsides now have no protection against the rains, and as a result they

▲ This lumber mill in British Columbia, Canada, is surrounded by rafts of logs floated down-river from the logging areas. Most of the wood-pulp used for paper-making and timber for building (in the Northern hemisphere) comes from softwoods, especially pines.

▶ A huge pile of dipterocarp logs awaits transportation from Sabah, Malaysia. The dipterocarp family, usually huge trees with tall straight trunks, dominates the rain forests of Asia and is the world's main source of hardwood timber (used for sawnwood, veneer and plywood manufacture).

are heavily eroded – their fragile topsoil washed away and carried into the lowlands.

THE TRADE IN TIMBER

About half the trees felled each year are used as fuel. Of the rest, about two-thirds are used mainly in the construction industry. The rest, used mainly for furniture, are among the most prized woods: teak, ebony, mahogany and rosewood.

In the natural forest such trees are widely scattered and mixed with scores of other species, many of which are not commercially valued. Even when the loggers try to cut these trees selectively, leaving the other trees intact, enormous damage is done. The tops of rain forest trees are often bound together with lianas (giant creepers), and when the target tree comes down it may bring several others with it. More damage is caused by the heavy machines that crush ground vegetation and tear deep gashes in the forest soil.

MANAGING THE RESOURCE

For many tropical nations timber is a major export and an essential source of income. The world demand for logs, sawn timber, and wood to be turned into plywood, blockboard, paper and cardboard is growing every year, and exporting nations have to earn a living. However, the forest resource could be used more wisely and could be far more efficiently managed. Numerous countries in South-east Asia have already taken steps to reduce their exports of rare timbers, and to build up their own furniture-making industries. That way the jobs associated with the timber stay in the wood-producing country where they are badly needed.

A DISASTROUS WASTE

Many of the importing countries are needlessly wasteful in their use of wood. Recent figures show that Japan now imports over 50 percent of all the tropical wood exported, yet a huge proportion of this is used for temporary timber-work, such as making concrete sections for high-rise buildings, and disposable chopsticks. The drain on the rain forests of Malaysia and Indonesia is so great that their natural forest has now almost gone.

In parts of North America and Europe, discarded paper and cardboard makes up over 40 percent of household waste. Most of it could be recycled, but two-thirds are thrown away. One research project in the USA showed that every ton of paper recycled would save 17 trees, 25 barrels of oil and 7,000 gallons of water. (In 1987, US consumption of paper was over 80 million tons.)

◀Amazonia's fires take only a few hours to destroy trees that have stood for hundreds of years. One satellite photograph showed 2,000 huge fires burning in the state of Rondônia alone. (Rondônia is about the size of the UK.)

▼A huge grassland fire in Queensland, Australia, burns where tropical rain forest has been cleared. One of the main fears today is that these clouds of smoke and gas from forest and grass fires are a major cause of the greenhouse effect.

POLLUTING THE WATERS

On 24 March, 1989, the tanker *Exxon Valdez* ran aground on Bligh Reef in Prince William Sound, Alaska. The jagged rocks tore a gaping hole in the ship's hull, and 35,000 tons (11 million gallons) of crude oil poured into the clear cold waters of this arctic inlet. Nearly 1,250mi of shoreline were fouled, and the death toll among local wildlife was estimated at 100,000 seabirds (including 150 Bald eagles), at least 1,000 Sea otters, and an unknown number of seals.

The *Exxon Valdez* spill was a harsh reminder of the enormous damage that can be done to coastal environments when mistakes are made in the handling of dangerous materials such as oil and toxic chemicals. An accident may happen in a matter of minutes, but the aftermath is often felt for years afterwards. No one should feel too easy about nature's ability to recover from such spills. The oil does not go away completely.

OIL AND OTHER POLLUTANTS

Although coastal tanker accidents make a huge local impact, the oil spilled from tanker accidents is small compared with the total amount entering seas from other sources such as polluted river outflows and discharges from shipping.

In 1970 Thor Heyerdahl, the Norwegian explorer, on his voyage across the Atlantic on the reed boat Ra II, found floating lumps of tar in the sea on three out of every four days of his voyage. Even then, in the middle of one of the world's largest oceans, the waters were polluted with oil from minor spills and the tanker-operators' practice (since outlawed) of washing out their tanks at sea.

Perhaps more important in the long term are other pollutants entering lakes, rivers and seas: sewage, waste products including heavy metals from factories, and agricultural fertilizers and pesticides washed from the land. This kind of pollution rarely makes the headlines, but year after year its effects build up – fouling the waters right across the world.

EXXON VALDEZ – THE RESCUE

In the months that followed the *Exxon Valdez* disaster a huge rescue and clean-up operation was mounted. Oil skimmers mopped up as much as they could of the oil slick floating offshore, while on land more than 11,000 workers attacked the sticky black mess that clung to rocks and coated beaches all along the coast. Using shovels and rakes and high-pressure water hoses the oil was scraped, shoveled and washed into piles, then loaded into plastic sacks for disposal. Some of the debris was burned in special clean-burning hospital incinerators. More than 50,000 tons of oil-drenched sand and gravel were later shipped to Oregon and buried in a toxic-waste dump.

Thousands of oiled birds, Sea otters and seals were collected in the wildlife rescue operation. Once soaked in oil, their feathers and fur pelts were no longer able to insulate them against the arctic cold. Quite apart from being unable to fly or swim, the oiled animals were in grave danger of dying of exposure. Each one was carefully cleaned, cared for, and finally released when fit enough to survive on its own.

THE BILLION-DOLLAR BILL

The cost of the clean-up itself was more than a thousand million dollars, but the costs did not end there. Prince William Sound is home to a thriving herring fishery and several major salmon hatcheries. Because of the fear of serious contamination of the plankton on which the fish feed, the herring fishing season was cancelled and the salmon catch reduced. All round the shores of the Sound and its many islands, teams of scientists collected samples of birdlife, fish, shellfish, seaweeds and other land and sea life in order to test for the effects of oil in the local food chains.

Prince William Sound will one day return to its former beauty. In some areas the recovery will be quick, in others slow. Winter storms will help disperse much of the remaining oil, while micro-organisms that occur naturally in the environment will also

play an important part in breaking down the scattered sticky residues.

THE SEWAGE PROBLEM

City engineers have estimated that for every million people living in a large modern city, the annual output of sewage is about 500,000 tons. Most of this biological waste, either in its crude state or partially treated, ends up in our rivers and in the relatively shallow waters around our shores. Add to this the enormous quantities

◀ Quite apart from the possible health hazards, one problem of discharging raw sewage at sea is that it immediately floats to the surface. Sewage consists mainly of fresh water, which is less dense than the surrounding sea water.

▼ In March 1978 one of the worst ever cases of marine pollution occurred when the *Amoco Cadiz* ran aground off the coast of Brittany. The entire cargo, over 220,000 tons of crude oil, was spilled, fouling the beaches of northwest France and killing thousands of seabirds (*inset*).

of biological waste material produced each year by the food processing industries and the scale of the problem becomes clear.

In seaside towns where sewage is discharged directly into the sea, the most obvious problem is often a dark stain spreading down-current from the end of an underwater outlet pipe. This unpleasant sight can be a serious drawback for a town that depends on tourism for a large part of its livelihood. Much worse, a shift in the local coastal currents, or a pipe that does not extend far enough from the shore, may well lead to sewage being washed or blown back on to the beaches causing a health hazard.

CHOKING THE RIVERS

Much more serious in the long term are the biological consequences of discharging organic waste into the confined waters of a stream or river. The high concentration of solid particles in untreated sewage prevents sunlight from penetrating beyond the surface layers. This stops the process of photosynthesis, killing off many of the water plants. Others may be smothered by the build-up of organic mud on the river bed and by huge growths of sewage fungus.

At the same time, the thick "soup" of fats, proteins and carbohydrates provides a flood of nutrients for micro-organisms in the water. These grow at an enormous rate, rapidly using up all the oxygen in the water. Soon, all the organisms that depend on oxygen disappear from the river. First to go are the fish, which are very sensitive to water pollution, but where the pollution is very bad even the most tolerant water plants are soon killed off by the build-up of toxic sulfides and ammonia.

POISONS IN THE FOOD CHAIN

One of the most serious forms of water pollution actually starts on land. In order to increase crop yields, farmers spray their fields with fertilizers, and with an ever-increasing number of herbicides (to kill off weeds and various fungi that attack crop plants) and insecticides (to control insect pests). These chemicals do their job, but eventually most of them are washed off the land by the rain, and into streams and rivers. The fertilizers provide yet more food for the organisms that choke the rivers, while the pesticides introduce a new and even more deadly threat.

The problem with many modern pesticides is that they are too good. They kill off the target pest, but when their job is done they do not go away.

◀ Despite all the odds, this Mute swan continues to nest on a heavily polluted lake. But its days here are numbered. Soon it will be forced to look for a new nesting site with clean water and uncontaminated food.

▼ If the waters of a lake contain just 0.00005 parts in a million of DDT, the concentration in water plants may be 800 times higher, in small fish 6,000 times, in the pike 33,000 times, and in the grebe over half a million times higher.

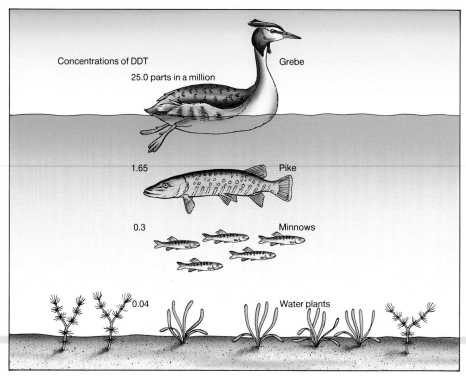

▲Using helicopters or light aircraft to spray crops with fertilizer or pesticide is fast and efficient, but if the chemicals are not sufficiently selective, and are blown onto nearby vegetation or farm animals, the results can be very damaging.

They are very persistent and remain in the environment for years. They also become absorbed into the food chain – with devastating results.

A pesticide may be sprayed onto the land at such low concentration that it is no danger to any but the target species. Even when it is washed into a river it may still be well below danger level. But then it is absorbed into the leaves and stems of water plants. These are eaten by plant-eating fish, which in turn are eaten by pike and other predators. Many chemicals are stored in animal body-fat, and so each animal takes in – and keeps – all the chemicals absorbed by the animals and plants it feeds on.

The build-up can be dramatic. A grebe or heron at the top of a river food chain may end up carrying a dose of pesticide *half a million times* more concentrated than that in the water.

BACK FROM THE BRINK

Among the worst offenders are the organochlorine pesticides such as DDT, which is converted into another form, DDE, in the bodies of animals. This chemical caused a disastrous fall in the numbers of Peregrine falcons, ospreys, sparrowhawks and Bald eagles in the 1960s. These birds of prey feed mainly on fish and on seed-eating birds, and they were found to have accumulated massive doses of the deadly DDE. Some died as a direct result of the poison, but many others simply failed to raise any young. One effect of DDT on Peregrine falcons was to cause the birds to lay thin-shelled eggs, which broke when the adult birds tried to incubate them. Only after DDT was banned from many countries did these birds begin to recover their numbers.

The deadly chemical was also found in the body-fat of Adelie penguins on the Ross Ice Shelf in Antarctica, thousands of miles away from the nearest possible source of the chemical. The doses were small, too small to cause serious harm to the birds, but the fact that it was there at all was dramatic proof of just how far and how quickly a persistent chemical could spread throughout the world's oceans.

THE HEAVY METAL THREAT

Heavy metals such as copper, zinc, lead, nickel, chromium, silver and mercury are used in many industrial processes. Most of these metals and their compounds are extremely poisonous. If they enter rivers they can do enormous damage to plant and animal life. Rainwater falling on the old spoil-heaps of abandoned metal-workings can create a hazard by washing metal residues into nearby streams.

Mercury is unquestionably the most dangerous heavy metal of all. Once in the human body it soon damages the brain and central nervous system. This can cause loss of all muscle coordination, followed by blindness, deformity and incurable mental illness.

THE MINAMATA TRAGEDY

One of the worst cases of mercury poisoning on record hit the Minamata Bay area of Japan in 1953. More than 2,000 people were affected altogether and about 400 died.

The cause of the Minimata disaster was discovered to have been mercury compounds discharged into the bay from a big vinyl chloride factory in Minamata City at the head of the bay. The compounds released into the bay in the factory waste were not soluble in water, but once they settled into the seabed mud they were converted into soluble compounds. These then found their way into the food chain. They were absorbed by shellfish and by fish – and finally by the people who ate the seafood caught in the bay.

▶ In the North Sea in 1987, this ocean incineration vessel *Vesta* burned and dumped in the sea a variety of toxic wastes.

POLLUTING THE LAND

In Hamburg, in 1979, an eight-year-old boy was killed and his two friends were badly hurt when chemicals left in an abandoned factory exploded while they were playing. Accident investigators found a time-bomb. The decaying factory buildings contained more than 60 different chemicals, including deadly nerve gases left over from the Second World War.

Germany is not alone. The nightmare of long-forgotten toxic waste dumps and the problems of dealing with today's dangerous wastes face all the industrialized nations.

THE INVISIBLE THREAT

In Britain, in 1980, over 4.1 million tons of hazardous industrial waste were dumped in the ground without any reliable safeguard against environmental damage. At many of these landfill sites, dangerous liquid chemicals were poured into the same hole in the ground as ordinary solid household refuse, and then buried in the hope that natural soil processes would neutralize the poisonous waste and make it safe.

In some cases this probably does happen. In others the toxic material does not break down: it remains unchanged, and over the years there is an ever-increasing danger that toxic chemicals will seep through the ground and contaminate soil and groundwater.

The scale of this invisible threat to public health became clear in 1980 when a case in the Netherlands hit the headlines. In Lekkerkerk, a suburb of Rotterdam, more than 270 homes had to be evacuated when the ground beneath a large housing estate was found to be contaminated with toluene, xylene and heavy metals. More than 900 people had to be re-housed and more than 150,000 tons of contaminated soil had to be dug out and disposed of by incineration.

THE CHEMICAL TIME-BOMB

Throughout Europe and America, and indeed throughout most of the world, there are now thousands upon thousands of these dangerous waste dumps. Some are relatively new, but many have been abandoned for years, and often their exact location is not even recorded. The deadly threat may lie hidden underground for decades, until some leak causes a tragedy – and a public outcry.

In 1981, several hundred tons of toxic solvents were discovered under a busy city square in the Danish capital Copenhagen. They had been discharged by a paint factory that had closed more than 10 years earlier. In France, some 23,000 cu.yd of oil waste leaked from a storage lagoon and contaminated the surrounding area. The lagoon had to be pumped out and the dangerous sludge shipped away for safe disposal. In Britain, during the 1960s and early 1970s a textile factory in Yorkshire dumped large quantities of asbestos waste near the village of Hebden Bridge. Medical investigators are still counting the cost in deaths from diseases directly linked to the dangerous asbestos fibers.

▲ The hills surrounding the settlement of Queenstown in Tasmania were once completely covered in vegetation. Now they are bare. The vegetation has been killed off by sulfur fumes from the local copper mine and processing plant, and the barren earth and rocks are stained yellow, gray, pink and purple.

▲ Not even the most remote parts of the world are free from the refuse of human occupation. Here at Point Barrow in Alaska a stretch of shoreline is buried under piles of garbage. Nothing can grow here, and as toxic material seeps into the water animal and plant life farther along the shore is affected.

WHAT IS BEING DONE?

These examples do not point a finger of blame at any particular country: they simply illustrate a widespread problem. What is surprising is how late the industrial nations were in waking up to the problem.

Regulations setting out safety standards for dumping, burning and other disposal methods were introduced in most countries only in the mid-1970s. Before that, toxic wastes were dumped almost anywhere, without controls and without proper records being kept. In fact, so little information was available on how much waste had been dumped, and where, that a German official summed up the situation by saying, "Ten years ago, most of Europe's toxic garbage vanished down a black hole somewhere."

The job of cleaning up Europe's old and abandoned waste dumps has now begun, but it will take years and the cost will be enormous. In the Netherlands alone, estimates range from US $3.2 billion to US $5.6 billion.

A WAY FORWARD?

Is it possible for dangerous wastes to be handled more safely in the future? The technology does exist.

Toxic and domestic waste should be separated, and the domestic waste burned or buried. Much of the toxic material can be disposed of safely in high-temperature incinerators. What cannot be burned should be buried – in carefully controlled landfill sites, in containers where necessary, and with a thick covering of compacted earth.

There are, however, two main problems. High-temperature incinerators are expensive to build, and proposals to open new landfill sites are always fiercely opposed by local residents.

AMERICA'S "SUPERFUND"

One way for Europe to tackle the problem of old waste dumps might be to follow America's example. In 1980, the US government established a fund to support the identification and clean-up of abandoned toxic waste sites. Between 1981 and 1985 the fund received more than US $1.6 billion. About 85 percent of the money was raised by a special tax on the raw materials used by the petrol and chemical industries. The rest came from grants from the federal government and from state governments.

The United States clean-up needs action on a huge scale. Estimates of just how many dumps will have to be made safe vary a great deal. At the lower end of the range the task could involve 2,500 sites and cost US $23 billion, with the clean-up taking 10 years. At the other extreme there are those who believe up to 10,000 sites may have to be dealt with – a job that could cost US $100 billion and take 50 years. In 1986 the Superfund was

strengthened. Priority for clean-up was focused on 375 of the worst sites, and alongside that work, money was set aside for research into more efficient disposal methods.

A BETTER FUTURE?

It is too early yet to know how far the American clean-up will succeed, or how quickly Europe will solve her own problems, but there are some positive signs. By 1983 most of the members of the Organization for Economic Cooperation and Development (OECD) – the main economic and trading "club," consisting of the European countries, Canada and the United States – had brought in (or were about to introduce) new systems to keep track of dangerous chemicals, from the transportation of raw materials right through the various industrial processes to the final disposal of waste products.

Many European countries are now making progress in the recycling of dangerous wastes, and destroying or burying the residue in carefully controlled landfill sites. Some countries, including Switzerland, France, Austria, Sweden, Denmark and the Netherlands have also opted for centralized systems of monitoring and controlling the movement and disposal of toxic waste.

One thing is certain. We cannot avoid dangerous wastes: they are a part of modern industry. We must, however, find ways of cutting down the amount we produce – and find better ways of recycling those that can be reused, and of destroying those that cannot. Burial should be a last resort, used only when there is no other means of disposal.

THE *KARIN B* SCANDAL

Few Third World countries have laws controlling production and disposal of toxic waste, and this makes them easy targets for those willing to dump their poisonous waste anywhere – so long as it is a long way away.

The *Karin B* affair started in Nigeria in 1988, when people living near the town of Koko became sick. The cause was found to be fumes coming from a huge dump of chemical drums near the town. There were 8,000 of them, badly corroded, rusting and leaking toxic chemical waste. The chemicals had come from factories in Italy, and had been dumped in Nigeria by an Italian-owned company based in the capital, Lagos.

The Nigerian Government insisted that Italy take back the drums, and eventually they were packed into containers and loaded onto ships. One, the *Karin B*, tried to return to Ravenna in Italy but the townspeople refused to let the ship enter harbor. She then tried Cadiz in Spain – and was turned away. The same happened in Britain, then in Belgium, West Germany, France and the Netherlands. Finally the Italian Government ordered the ships to sail back to Italy and to unload at a military dockyard. There the waste was off-loaded into a specially built storage facility to await proper disposal.

The problem arose because Italy had too few of the high-temperature incinerators needed to dispose of the toxic waste produced by her industries. Someone came up with the idea of exporting some of it. In this case the plan did not work, but the long voyage of the *Karin B* and her foul cargo shows the scale of the problem of toxic waste – and the lengths people will go to in order to get rid of it.

◀ Disposal of liquid high-level radioactive waste from nuclear power stations is a major problem: it cannot simply be thrown away. Here at Hanford, in Washington, USA, the first one-million gallon long-term storage tanks are being built. Each tank will have a double wall of carbon-steel and the space between the walls will be constantly monitored for any sign of leakage from the inner container.

POLLUTING THE AIR

Throughout the summer of 1986 Robert Bruck of Carolina State University measured the levels of air pollution on top of Mount Mitchell in North Carolina. He also measured the rate at which the mountain's fir and spruce trees were losing their needles. His records showed pollution far above government safety levels. All around him, the forests were dying under a rain of industrial pollution.

Two main kinds of pollution affect the air we breathe. The first consists of solid material – tiny particles of dust and smoke that are carried into the air from factory chimneys, forest fires, agricultural land and vehicle exhausts. The second is made up of invisible gases produced by power stations, industrial and chemical processes, and millions of car and truck engines.

THE GREAT TURNING-POINT
There has always been some solid material floating around in the air. Pollen from plants and spores from fungi are found high in the atmosphere. So too are salt crystals from sea spray, particles of soot from forest and grassland fires, and rock dust from the arid regions of the world.

Most of this material is too fine to see, but not all. At least once in most summers, car owners across Europe wake in the morning to find their parked vehicles covered in a film of fine pink dust that has been blown north from the Sahara Desert.

Until the mid-19th century the level of these solid pollutants was not a problem. Everything was to change, however, with the Industrial Revolution. To power the new factories, steam was generated by coal-fired boilers which poured thick smoke into the air. Iron and steel mills, and a host of other industries, added their own clouds of smoke and dust. The internal combustion engine added yet another source of soot and gas as petrol and diesel vehicles began to multiply throughout the world.

We have only recently realized that the Industrial Revolution was not just a technological turning point in history: it was a turning point in our relationship with the environment.

HOW PLANTS ARE AFFECTED
The most obvious effect of atmospheric dust can be seen on plants. The dust forms a film over the leaves, and this has two important effects. First, it blocks out sunlight, which reduces the plant's ability to carry out photosynthesis, and secondly it interferes

▼Lichens vary enormously in their reaction to pollution. Some thrive in polluted air: others hate it. In just 20 years, between 1953 and 1973, the very tolerant *Lecanora conizaeoides* almost doubled its range in the British Isles, while the sensitive *Lobaria pulmonaria* shrank to only half its former range.

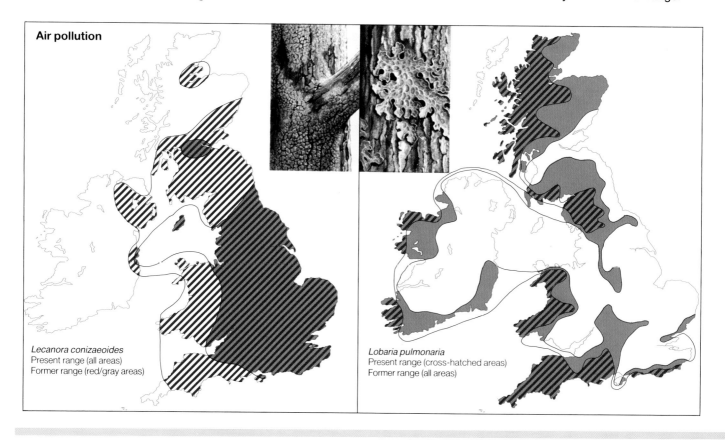

Air pollution

Lecanora conizaeoides
Present range (all areas)
Former range (red/gray areas)

Lobaria pulmonaria
Present range (cross-hatched areas)
Former range (all areas)

▲ In Tokyo, Japan, pollution of the air by vehicle exhaust fumes can become so severe that traffic police and some pedestrians wear filter masks when they venture onto the streets.

with the opening and closing of the leaves' pores – tiny holes that allow water vapor, carbon dioxide and oxygen in and out (see page 12). Coniferous trees such as Scots pine and Norway spruce are particularly badly affected by this type of pollution.

The gases most dangerous to plant life are sulfur dioxide, all fluorides, ozone and the oxides of nitrogen, all of which can damage vegetation many miles away from the source of the pollution. Also highly damaging, but more localized, are hydrocarbons, and nitric and hydrochloric acid. Most of these poisonous gases are created by power stations and car engines, and by manufacturing industries such as steel mills, chemical works and refineries.

Sulfur dioxide and nitrogen dioxide in the air damage plants by withering their leaves and turning them yellow. The gases are also dissolved in rainwater which seeps into the ground and is taken up by plant roots, which restricts their growth.

Trees, cereals, fruit crops and vegetables are all affected – and the cost in lost farm and timber production can be huge. Toxic fumes from a copper-smelter at Redding in California damaged natural vegetation and orchards over an area of 20sq mi, while the gas pouring from another smelter at Washoe in Montana destroyed timber plantations more than 20mi away. Douglas fir, Lodgepole pine, Subalpine fir and Timber pine were all badly affected.

TELL-TALE LICHENS
Different groups of plants react in very different ways to pollution. Some thrive in polluted air, others hate it. Some are so sensitive that they can even be used as indicators of the level of pollution.

Some lichens that grow on trees and rocks are particularly sensitive to sulfur dioxide gas, and scientists have devised a scale that relates the abundance of *Lecanora* and *Parmelia* lichens in England and Wales to the amount of sulfur dioxide in the air.

URBAN AIR POLLUTION
Even as recently as 30 to 40 years ago, many cities were plagued with smog – an unhealthy mixture of natural fog and industrial smoke that could hang over a city for days, often causing hundred of deaths from bronchitis.

In recent years many of these cities have cleaned up their air by establishing "smokeless zones," and by enforcing limits on the amount and type of smoke and gas that can be emitted from factories. Where cities lay in valleys or were surrounded by hills, factories were required to build taller chimneys so that the killer smoke would not become trapped over inhabited areas. That also helped, at least locally, but pollutants pumped high into the air were soon to produce a new threat – acid rain.

DANGER FROM CAR EXHAUSTS

Vehicle exhaust is a serious risk to health in many major cities. In Tokyo air pollution often becomes so severe that police officers on traffic-control duty wear face-masks to filter out some of the poisonous fumes.

There are now so many vehicles crammed onto city roads that the air is often blue with fumes. Many countries now have regulations to limit exhaust emissions from new vehicles, but car numbers increase so fast that overall pollution still worsens.

Los Angeles in California suffers one of the worst forms of car-linked air pollution. There, the shape of the land and the local atmospheric conditions prevent car exhaust gases from dispersing. On still days they hang over the city, and the ultraviolet light in the Californian sunshine causes chemical reactions to take place in the exhaust fumes. These produce a very dense haze of "photochemical smog," a choking mixture of pollutants that contains highly toxic peroxyacetyl nitrate (PAN), along with formaldehyde, ozone and nitric acid.

HOW TO LIVE WITH THE CAR

The two quickest and most effective ways of reducing nitrogen oxides and hydrocarbons in the air would be to limit speed and insist that all cars be fitted with catalytic converters. At high speeds, most car engines are inefficient: their exhaust contains large amounts of hydrocarbons. Reducing speed limits would keep most cars running closer to their peak efficiency. This in turn would reduce fuel consumption and emissions. Catalytic converters clean up the exhaust even further by removing any unburned fuel it contains.

One particular problem with car exhaust is the lead that is added to petrol to prevent engine "knock." About 75 percent of the additive is pumped back into the atmosphere in the exhaust gases, and there is growing evidence that this heavy metal is a serious health risk, especially to children. The result has been a trend toward lead-free fuels.

ACID RAIN

Sulfur dioxide and all the oxides of nitrogen released high in the air from factory chimneys drift far and wide on high-level winds. As they do so they also undergo chemical changes. Ozone in the upper atmosphere reacts with the gases to produce sulfates and nitrates, and these are then converted into dilute sulfuric and nitric acid. Hundreds of miles downwind from the original source, acid rain falls on forests and fields causing extensive damage.

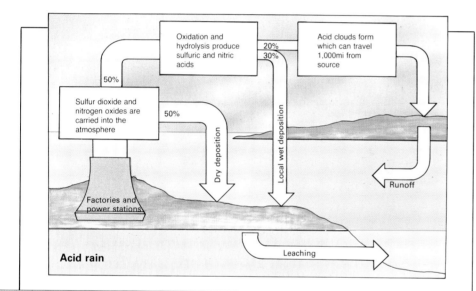

◀▲ Acid rain

Sulfur dioxide and nitrogen oxides from factory and power station chimneys may fall to the ground as tiny solid particles ("dry deposition") or as acid rain (or acid snow) after being dissolved in atmospheric water. The acids can damage plant life, kill fish and other life in lakes and rivers, and leach nutrients from soils.

The map on the left shows the acidity of precipitation over North America. The pH value of rain and snow was monitored at sites all over the continent, and the map shows areas with the same acidity value. Lakes in areas with carbonate rocks can counteract the acid to some extent, but lakes lying on acid rocks such as granite have no natural means of resisting this pollution.

pH values
lower values denote higher acidity

4·1
4·2
4·4
4·6 — Eel, brook trout die
5·0 — Perch, pike die
5·5 — Whitefish, siklöja, grayling die
6·0 — Sensitive insects, plant and animal plankton, salmon, char, trout, roach die
Crustaceans, snails, molluscs die

Industrial areas

Acidity is measured on a scale of pH values, and normal rainwater has a pH value of 5 or just above. Across large areas of eastern North America, Scandinavia and Germany, rainwater now has an average pH of 4.0 to 4.5. A decrease of one unit on the pH scale means a *tenfold increase* in acidity.

Apart from vegetation, acid rain affects lakes and rivers too, with devastating effects on aquatic life, especially on fish such as trout and salmon. In Sweden, more than 2,500 lakes have suffered extensive damage

▼▶ Acid rain is a complex issue. In some areas forests are damaged directly by acid rain and snow. However, in some parts of Germany where forests were suffering badly, scientists found that the forests were still rich in lichens that were very sensitive to sulfur pollution. It seems that part of the problem may be increased levels of damaging ozone, produced by the action of sunlight on motor vehicle exhaust gases.

One source of carbon dioxide in the atmosphere is the constant burning-off that takes place in oilfields as a means of removing unwanted hydrocarbon gases. In the Gulf War of 1991, this form of pollution was greatly increased by the willful setting on fire of oilfields in Kuwait by Iraqi forces. The global warming that results may alter the climatic zones, vegetation and farming zones.

to their fish populations, and another 6,500 have acidity levels way above normal. In Norway, more than 1,750 lakes are without fish and 900 more are known to be in danger.

THE GREENHOUSE EFFECT

More than 40 percent of the radiation reaching the Earth from the Sun is short-wave radiation which passes straight through the atmosphere. This radiation warms the ground, and the Earth then radiates energy back into space – at longer wavelengths. Most gases allow this longer wavelength radiation to pass straight through as well, but carbon dioxide in the atmosphere absorbs it. The carbon dioxide acts as a heat trap by allowing radiation in and then preventing it escaping again, just like the glass in a greenhouse. The result is a gradual raising of global temperatures.

The problem today is that human activities release millions of tons of carbon dioxide into the air every year

▼Carbon dioxide, water vapor, methane and a number of other gases that are present naturally, or produced by human industrial activity, absorb long-wave radiation from the Earth's surface and then re-radiate it back to Earth. This "thermal blanket" will cause global warming. Should this become serious, the polar ice caps may melt, causing flooding of low-lying coasts.

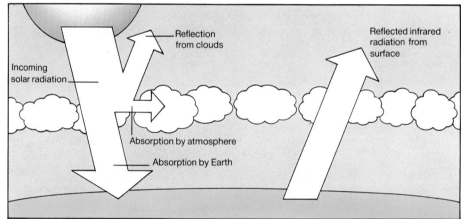

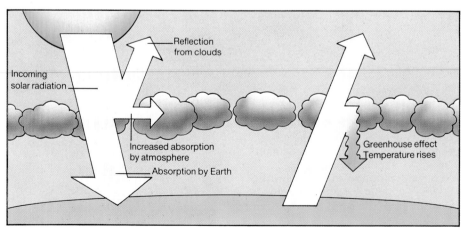

– from car exhausts, factory chimneys and forest fires. Methane, nitrous oxide, and the chlorofluorocarbons (CFCs) used in aerosols and refrigerators, all have the same effect. They are present in much smaller quantities but add to the general "thermal blanket" of the greenhouse gases (GHGs).

Scientists are still trying to predict the effects of this global warming. Estimates vary, but even a small rise in average temperatures could alter the pattern of climatic belts, making continental interiors drier and coastal areas wetter. This in turn could drastically alter world food production. A larger rise in temperature, say of 7°F, might trigger the melting of the polar ice caps. This could raise sea level by 160ft, which would flood many of the world's most densely populated low-lying coastal regions.

THE HOLE IN THE OZONE LAYER

Animals and plants are protected from the harmful effects of ultraviolet (UV) radiation by a layer of ozone gas 6 to 30mi high above the Earth. The ozone itself is unstable: it is constantly forming, breaking down and reforming. In the 1970s there was concern that gas emissions from high-flying supersonic aircraft might damage the protective layer permanently. There were also fears that the ozone layer might be damaged by increasing use of agricultural fertilizers, and by nuclear weapons testing.

These fears proved groundless, but in the mid-1970s a new threat was discovered. At ground level CFC gases are very stable, but once they have drifted into the upper atmosphere they are broken down by UV radiation. The chlorine that is released reacts with atoms of ozone to form other compounds, and this reduces the thickness of the ozone layer. Scientists have discovered a very marked thinning of the ozone layer over Antarctica, and because of the threat of long-term damage the use of CFCs has been banned in some countries and restricted in many others.

THE CHERNOBYL DISASTER

On 26 April 1986 a nuclear reactor at Chernobyl in the Soviet Union exploded and caught fire. Radioactive debris was blasted 6,500ft into the air, and within minutes a huge plume of contaminated smoke and dust began to drift across large parts of the Soviet Union and north-west Europe. At the time only 31 deaths were reported, mainly among the fire-fighting crews, although several hundred people were hospitalized with radiation sickness. The nearby town of Pripyat was evacuated, and will remain a ghost town, heavily contaminated, for many years. What the final toll will be is uncertain, but estimates by medical officers suggest that over the next 30 to 60 years between 5,000 and 50,000 people could die from cancers caused by the nuclear accident.

Chernobyl became a household name overnight. Throughout Europe and far beyond, governments set up teams of scientists and engineers to review safety procedures. Sheep and cattle grazing on land contaminated by the fall-out cloud had to be destroyed, and many upland grazing moors are still "out of bounds" due to lingering radioactive contamination.

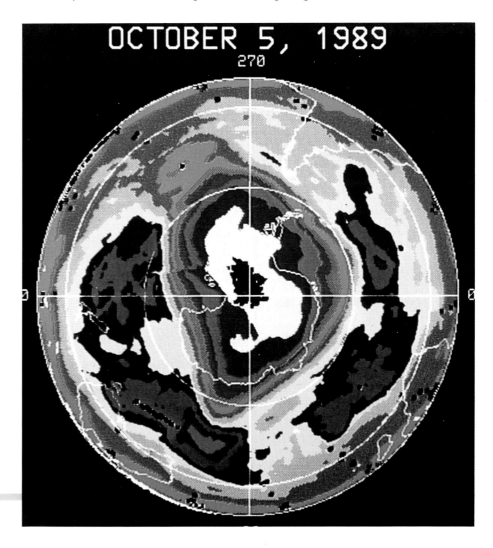

▶ This map shows the hole in the ozone layer over Antarctica. The thinning of the ozone layer is probably due to pollution of the upper atmosphere by chlorofluorocarbons (CFCs) widely used in aerosols and refrigerators. In many countries use of CFCs has now been banned.

PESTS AND PEST CONTROL

The Colorado beetle's original home was North America's Rocky Mountains. The insect itself was quite rare because its natural food, a relative of the potato, was scarce. When European settlers spread across North America they brought with them new crops, including potatoes, and very soon this "harmless" beetle became a major crop pest.

▼ The Brown rat (*Rattus norvegicus*) is found all over the world around human habitations. It causes damage to food and property and can spread disease.

The Colorado beetle story illustrates two of the ways in which a relatively harmless species can become a major nuisance. When a new crop is introduced to an area it can provide an abundant source of food for animal species that had been controlled by limited natural food supplies.

NATURE'S OPPORTUNISTS

Many of these animals are opportunists, and as soon as they "discover" the new food source they spread rapidly and reproduce at an enormous rate. In many ways they are like pioneer plants (page 32), taking advantage of any suitable new niche that becomes available.

A second factor is that when an animal is moved to a new area it often leaves behind its natural enemies. Many insects are kept in check by their natural predators, and so once they are free of them they multiply very rapidly.

WHAT IS A PEST?

In the natural world there is no such thing as a pest. The plants we call weeds and the insects we brand as pests are simply very successful members of the plant and animal community. It is their success as species that brings them into conflict with people. They are just too successful for their own good.

◄▲ The smallest insects can do untold damage to food crops. Aphids, for example, damage cereal crops both by sucking the plants' sap and by introducing viruses like the barley yellow dwarf virus illustrated above. To reduce the amount of pesticide used, biologists try to control the pests by encouraging predatory beetles (*left*) and wasps.

▲ The devastation caused by locusts is hard to imagine. For part of their lives these insects hop around on the ground in small numbers, causing little damage, but after breeding they take to the air in dense clouds containing thousands of millions of insects. Here, a swarm of Australian plague locusts has stripped an entire pea crop to bare stalks.

THE DISEASE CARRIERS

A "pest" is an animal or plant that does some harm to people. This may be direct harm such as in the case of malaria, sleeping sickness and other diseases spread by biting insects. It may also take the form of illnesses caused by parasites that live in the human body. Some are caused by parasites whose eggs have been swallowed in polluted drinking water. Others may be picked up in cuts on the hands or feet. The female Human bot fly lays her eggs directly on human skin and the larvae burrow into the flesh causing festering wounds.

CROP PESTS

The other main types of pest are those that do economic harm. The most obvious are crop pests – leaf-eating insects, stem and root borers, birds such as Wood pigeons which eat huge quantities of seeds and young plants (notably cabbage and cauliflower), and the rats and mice that do enormous damage to grain crops.

Fungi, too, are a major cause of damage, both to growing plants and to stored crops. Potato blight can ruin entire crops of potatoes and tomatoes while a variety of molds will attack virtually any kind of stored organic material from maize and barley to wool, flax, leather, paper and wood.

THE ECONOMIC COST

All our food comes from plants, either in the form of food crops or through the grazing land, fodder crops and food supplements on which we feed our farm animals. The total cost of pest losses every year is huge. Over the whole world, roughly a third of the annual harvest is lost – either in the field or through rot, fungus and pest damage to food and other crops in storage. In some countries the losses are even greater. In Latin America, for example, more than 40 percent of all agricultural produce is lost.

Even in the United States, where farmers have all the benefits of advanced technology, pesticides and fungicides, nearly 30 percent of the annual harvest is lost, at a cost of US $20,000,000,000 a year.

NON-FOOD CROPS

Quite apart from the damage to food crops, pests cause enormous damage to many other commercial products. Cotton, sisal and jute grown for the manufacture of cloth and rope are attacked by numerous pests. The Food and Agriculture Organization of the United Nations (FAO) has estimated that without pesticides half the cotton crop in the developing countries would be ruined.

The problem is not confined to the tropics. In temperate North America, occasional outbreaks of Spruce bud moths kill millions of trees, including valuable Balsam firs and White spruce.

PESTS IN THE HOME

Even in towns and cities insect and fungus pests invade our homes and damage property. Several kinds of beetle that normally inhabit dead and dying trees have through the ages found timber floors and rafters to their liking. One survey showed that in southern Britain at least 85 percent of houses more than 50 years old had some woodworm damage.

In tropical regions the problem can be even more serious. Termites and fungi together can destroy an entire wooden house in 5 or 6 years.

There are nuisance pests too. Flies and cockroaches contaminate food and can spread disease. Moth larvae damage stored clothing, while carpet beetles and museum beetles ruin fur and leather.

PEST CONTROL

The principles of pest control are quite simple: they aim to prevent the pest reaching its target, kill it (or at least reduce its numbers) using a pesticide, or control its population with a biological agent such as a natural predator.

Many people in the tropical regions sleep under mosquito nets and fit fine mesh fly screens over doors and windows to avoid being bitten and infected by disease-carrying insects. Although it is expensive, crops too can be protected by barriers such as greenhouses and bird netting. To keep out rats and other pests, produce where possible can be stored in silos and closed bins.

CHEMICAL WARFARE

The most widespread answer to pests is the use of chemical warfare. Among the most effective man-made insecticides are the organochlorides, a group that includes DDT, dieldrin and lindane (see page 63). At first they were thought to be the answer to many pest problems. They were widely used in agriculture worldwide in the 1940s and 1950s. They are powerful, and very persistent – that is, they remain active for a long time.

What was not known at the time was that DDT is stored in animal fat, and it soon started to accumulate in natural food chains, with serious consequences for many top predators. Once the problem was discovered, the use of DDT was banned in many countries and the search began for pesticides that would not have such damaging side-effects.

The replacements were the organophosphate compounds – now the most widely used insecticides. They are extremely toxic to insects, but as they are short-lived they do not stay long in the food chain and are therefore much safer.

CHEMICAL BULLETS

Some new man-made compounds can be "aimed" at particular pest species. They are soluble in water, harmless to plants, and can be sprayed onto crops, which take in the chemical through their roots. The pesticide is then carried in the sap to all parts of the plant, with lethal effects on insects that feed on them.

BIOLOGICAL CONTROLS

Using natural predators as a form of pest control was first tried in California over 100 years ago. There, the valuable citrus orchards were plagued with Cottony cushion scale insects. When a species of ladybird was imported from Australia the effect was dramatic and in a few years the scale insects were no longer a problem. The natural balance of predator and prey kept the numbers of scale insects below the nuisance level.

The same approach was successful in Australia where the Prickly pear cactus, introduced as an ornamental plant, invaded vast areas of pasture. An introduced moth whose larvae fed on it brought it under control.

Today there is great interest in this "back to nature" method of control. It must be used with care, because introducing any alien organism into a new environment can be dangerous. The newcomers may get out of control, or have unexpected effects on the local food-chains.

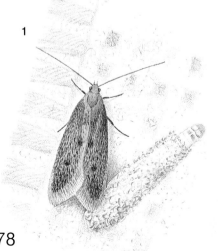

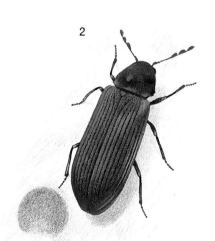

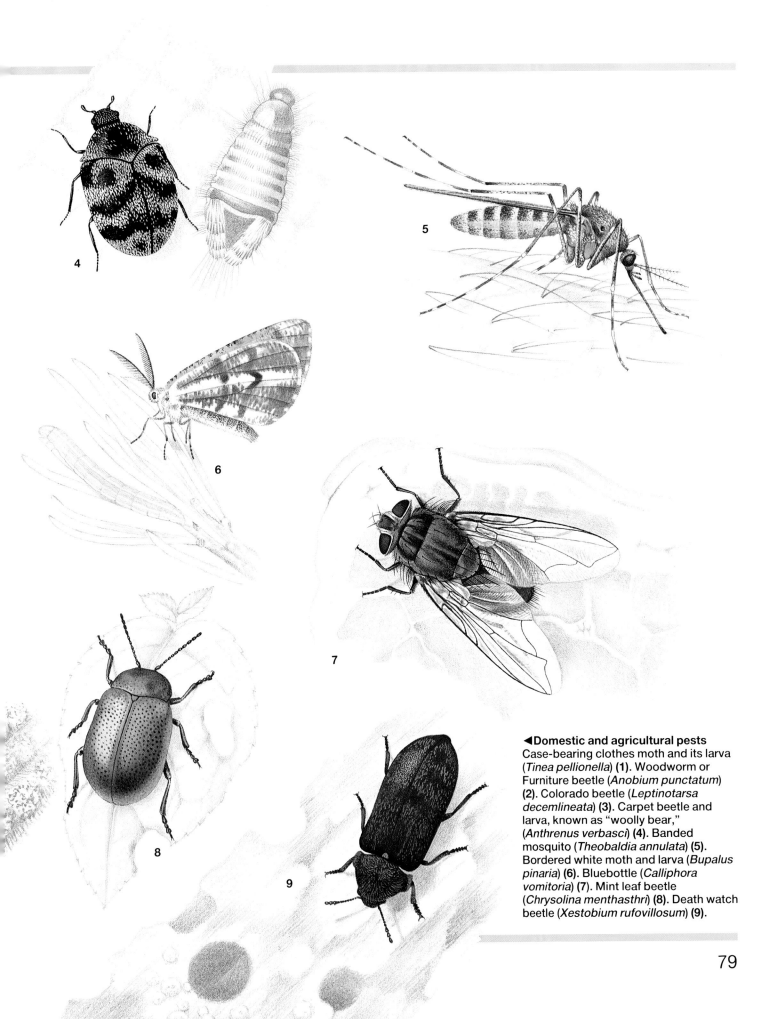

◀**Domestic and agricultural pests**
Case-bearing clothes moth and its larva (*Tinea pellionella*) **(1)**. Woodworm or Furniture beetle (*Anobium punctatum*) **(2)**. Colorado beetle (*Leptinotarsa decemlineata*) **(3)**. Carpet beetle and larva, known as "woolly bear," (*Anthrenus verbasci*) **(4)**. Banded mosquito (*Theobaldia annulata*) **(5)**. Bordered white moth and larva (*Bupalus pinaria*) **(6)**. Bluebottle (*Calliphora vomitoria*) **(7)**. Mint leaf beetle (*Chrysolina menthasthri*) **(8)**. Death watch beetle (*Xestobium rufovillosum*) **(9)**.

SHARING PLANET EARTH

At one of South Africa's national parks, game wardens were surprised when the number of waterbuck fell alarmingly. At first they thought that too many were being killed by lions, but further investigation showed they were being pushed out of their usual grazing areas by another antelope, the nyala. The waterbuck, forced to feed in other areas, were dying from diseases carried by ticks. It was not predation that caused their decline but competition. When park managers removed some of the nyala, the waterbuck returned and quickly recovered.

One of the main reasons why it is necessary to manage habitats for conservation purposes rather than simply leaving them entirely alone is that the habitats themselves do not remain unchanged. Slowly, over a period of time, the vegetation will change (see the section on successions, pages 32 to 37). As the vegetation changes, so too will the animals that inhabit the area. Some may find their preferred food plants or prey species have disappeared. Others may find they can no longer find the kind of burrow or nest sites they prefer. These animals will move on, and will be replaced by others who find the new conditions suit their particular life-styles.

HABITAT MANAGEMENT
Apart from a few of the largest national parks of Africa, Asia, North America and Australia, most protected areas are relatively small, several of them as small as a few acres. Within such a small area the natural processes of change can completely alter a habitat in a very short time. If these smaller reserves are to succeed in their aim of protecting and preserving examples of different habitats and the animals living in them, some form of management is usually essential.

POSITIVE INTERFERENCE
Conservationists often have to step in and either prevent or change natural processes, even though this may at first seem like interfering with nature. For example, if a stream feeding a

▶ Norris Geyser Basin in the Yellowstone National Park, USA. Almost one-third of the United States area is given over to National Parks, nature reserves and other managed lands. Human leisure activities that do not endanger the local environment are often allowed.

▼ Wicken Fen Nature Reserve in England. Wetland habitats such as this fen and reed marshland are home to a great variety of plants, insects and birdlife. It is important that they are not drained to make way for development.

◀The eland is usually found in open woodland or grassland, but unlike many of its relatives it is also found in scrub and semi-arid areas. This handsome spiral-horned antelope is under threat in some areas from agricultural development, habitat destruction and illegal hunting.

freshwater marsh were to alter its course, as streams often do, the marsh would very quickly begin to dry. The marsh plants would soon be replaced with dry-land grasses and shrubs, and the fish, snails, frogs, insects and birds of the marsh would disappear. It follows that if that stretch of marsh was turned into a nature reserve, one essential task for the managers would be to ensure that the stream kept flowing – even if that did then mean "interfering" with nature.

DIFFICULT CHOICES

Conservationists must sometimes make difficult decisions about the kind of management to adopt. Take the case of the Red-billed chough, a relative of the crow, which once lived on sea cliffs around the shores of western Britain and Ireland. The reason the birds have become so rare lies in local farming practices.

The cliff-tops were originally used for sheep-grazing, and this kept the grass short, which is the way the choughs prefer it. They feed on insects in the grass and prefer it short so they can easily see any approaching danger. But in some areas the cliffs are no longer grazed, and the grass has grown. To bring back the choughs, the best management system would be to cut the grass.

But what should be done if a particular cliff-top is home to a rare butterfly? Insects naturally prefer long grass, and cutting it might wipe out a whole local population of that species. Sometimes there is no easy answer. In a case like this the conservationists must look at how many choughs there are altogether, how

many other nesting cliffs are available to them, how many populations of the butterfly there are and how many places they have to breed.

CULLING TO PROTECT SPECIES

Another very difficult question for conservationists is whether or not animal populations should be controlled by culling – that is, by killing a certain number of animals when the population grows too large for the available food supply. Most agree that it is necessary, but it is a grim task, and one that rarely gets much sympathy from the public.

Elephants are immensely popular and the idea of shooting them, even "for their own good," is hard for many people to accept – especially as in some areas up to about 90 percent of elephants have been wiped out by ivory poachers in the last 10 years.

The hard fact is, however, that an adult African savannah elephant needs 330lb of food a day, and if the elephant population in an area becomes too large, the animals can completely devastate the habitat.

In the rainy season the elephants feed mainly on grass, but in the dry season they eat the leaves, twigs and bark of trees and bushes. Once its bark has been stripped off, a tree will die and if this kind of damage is widespread the habitat itself is on a downward spiral of destruction. There is really little choice but to cull if the savannah vegetation of such an area is to survive.

TO INTERVENE OR NOT?

Occasionally conservationists are faced with a habitat problem in which it seems better not to take any direct action. Take the Great Barrier Reef for example. This enormous natural reef complex runs for more than 1,260mi down the north-eastern coast of Queensland, Australia. It is home to a spectacular variety of coral species and a wealth of animal life that is one of the wonders of the natural world. Unfortunately, in recent years this unique reef has been under constant attack from a plague of predatory Crown of thorns starfish which feed on the living coral organisms. The result has been the death of large sections of the reef.

Some scientists argued that an attempt should be made to clear the reef of these voracious predators, or at least control their numbers. Others argued very strongly that in this case the reef might be the best manager of its own affairs. It is not just a section of habitat that has been declared a national park: it is an entire ecosystem in its own right.

For hundreds of thousands of years the reef has survived and grown. Scores of species must have come and gone in that time, and many conservationists believed that other plagues must have hit the reef in the past – but the reef has survived. To intervene on the scale that would be necessary to remove the Crown of thorns might do more harm than good to this ancient ecosystem.

▲ A Crown of thorns starfish moves over the surface of a living coral, feeding on the coral polyps – the tiny creatures that have built the Great Barrier Reef. In some parts of the reef more than 15 starfish have been found per square yard of reef, and scientists believe that population explosions like this may happen every 70 years or so.

IS CONSERVATION NECESSARY?

There are two main arguments for the conservation of habitats. One is that humans are the most powerful species on Earth and therefore have a

▲ In some parts of South Africa, efforts at conservation have been so successful that populations of some species are now too high for the carrying capacity of the habitat. These "surplus" White rhinos can be used to stock areas where local populations are still low.

▼ Human activities sometimes produce unexpected benefits to wildlife. Inside the perimeter of the Fermilab research site west of Chicago, native plants and animals, including rare Yellow-winged grasshoppers, are re-creating the old habitat of the American prairie.

responsibility to be the "caretakers" of our planet's wild creatures and wild places. After all, many of the greatest threats to nature come from mankind in the first place. We have the knowledge, the ability and the technology to save and preserve a great deal of what is now in danger. The big question is whether or not we are prepared to commit the huge amount of effort and money required.

The second main argument is that for our own good we should take care of the natural resources of the Earth. We depend on them for survival just as much as the plants and animals do, and if we fail to take care of them we, as well as the rest of the planet's inhabitants, will suffer in the long run.

USEFUL PLANTS AT RISK

Living in the relative comfort of the developed world we often overlook how much we owe to the distant wild places of the Earth. Most of our food comes from about 20 main crop plants, and the majority of these originated in the tropics where habitat destruction is now most severe.

Maize, cotton, potato, rubber and cocoa are all South American plants; banana, coconut, rice and sugar cane came originally from South-east Asia; oil palm, millet and coffee are from Africa, and so on.

The crop plants we rely on have been improved over the years by selective breeding. They are now bigger, stronger, more resistant to disease and produce far more food than their wild ancestors. But many of them can very likely be improved still further by "transplanting" useful genes from other wild plants.

Quite recently a young South American botanist discovered just such a "gene resource" in the mountains of Mexico. The new species is a wild relative of corn, and the discovery caused great excitement. This is because the plant is a perennial – one that will grow afresh year after year – while most cereals must be planted from seed to produce each crop. The new species also appears to be naturally resistant to many of the diseases that affect corn crops.

This new plant could have an enormous impact on world food supplies if the perennial gene and the disease resistant genes can be bred into high-yield crop plants.

ANCESTRAL ANIMALS

If we are to improve our food crops and animals we must preserve their wild ancestors and relatives. Domestic cattle, for example, are descended from the aurochs, which became extinct in 1627. There are now only a few wild cattle left – the banteng of South-east Asia's tropical forests, the gaur of India's forests, the yak of the Tibetan mountains and the kouprey which lives in the wooded savannahs of Indo-China. Of these four, two are listed as "Endangered," the other two as "Vulnerable." Yet these are the only surviving close relatives of the aurochs – the only ones that carry the ancestral cattle genes.

National Parks of South Africa
Of South Africa's 11 National Parks, only the Kruger Park and the Kalahari Gemsbok Park are large enough to contain the full range of plants and animals that live in their local areas. Yet even the one-million-acre Kruger National Park requires strict management to ensure a balance between the resident animals and the available food supply.

▼A giraffe in the Kruger National Park where windmills are now used to pump water from deep underground to help maintain the health of the habitat.

▲South Africa led the African continent in adopting conservation policies. The Umfolozi Game Reserve was established in 1897, and the Sabie Game Reserve – later enlarged and renamed the Kruger National Park – was opened in 1898. The country now has over 20 protected areas, ranging from the sun-scorched Kalahari Desert, through the rugged landscape of the Royal Natal National Park, to the southern coast Wilderness Lakes.

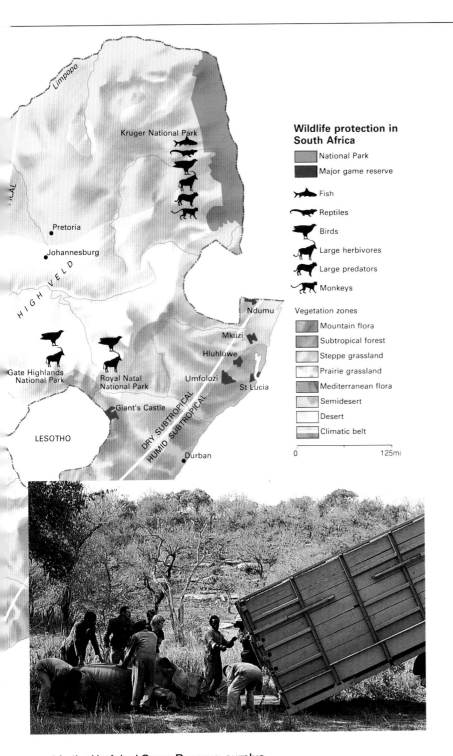

Wildlife protection in South Africa

▲ Stolsneck Dam, in the Kruger National Park, caused a considerable degree of environmental damage by flooding a large area of land, but it also brought benefits to the region. The lake provides a watering place for numerous game species, and a breeding place for fish and waterbirds.

▲ In the Umfolozi Game Reserve, surplus White rhinos are caught using tranquillizer darts fired from chase vehicles. They are then loaded into trucks and transported to other reserves where there is space to support them.

LIFE-SAVING PLANTS

In the United States in the mid-1970s doctors wrote more than 1.6 billion prescriptions a year. Nearly one-fifth of these were based on just five chemical compounds obtained from tropical forest plants. We are destroying the tropical forests at a staggering rate – and with them we are wiping out countless plant species before they can even be studied. Many of them could be life-savers.

The Rosy periwinkle, a native plant of Madagascar, is the source of two drugs that have proved enormously effective in the treatment of leukemia in children. Tropical forest plants have also given us quinine, morphine and many other drugs. How many more might still await discovery?

WILDLIFE CONSERVATION

In the summer of 1985 a young White-tailed sea eagle flapped its huge splay-tipped wings and launched itself into flight from a rocky cliff on the Scottish island of Rhum. For bird conservationists in Britain and Norway it was a moment of triumph. The eagle had not bred in its ancient Scottish home for more than 70 years: now, after a 10-year effort by scientists, the White-tail once more soared over the Scottish Highlands.

The awe-inspiring White-tailed sea eagle is one of many birds of prey to have suffered at the hands of people. The birds are opportunists, and as well as preying on fish and sea-ducks they will just as readily feed on carrion such as dead sheep, rabbits and deer.

Like many carrion feeders these eagles were branded as sheep-killers, and in many of their homelands, including Britain, they were wiped out by sheep-farmers in the late 18th and early 19th centuries, mainly by shooting and poisoning.

PATIENCE REWARDED
In 1976, bird conservationists from Britain and Norway began a joint attempt to reintroduce the eagle to its former Scottish home. Over a 10-year period over 80 young eagles were collected in Norway, where the species was still quite numerous. They were flown to Britain and released on Rhum in the hope that they would settle there and eventually breed when they reached maturity. There were many disappointments, but in 1985 four pairs mated and produced eggs. Three pairs failed to raise any young, but the fourth succeeded and the first native-born White-tail for over 70 years took flight over the wild coast of western Scotland.

▲ Hedgerow plants like the Garlic mustard (*above*) support the Orange-tip butterfly. But between 1945 and 1972, 80 percent of Britain's hedgerows were torn up to make way for bigger fields. Saving the rest is an urgent priority because hedgerows are home to birds, insects, mice, reptiles and many other species.

SAVING THE BLACK ROBIN
A number of other bird species have also been taken from one area and successfully resettled in a new home. One is the tiny Chatham Island black robin. At one point the population of this small New Zealand bird fell to only seven individuals – the smallest known population of any endangered bird. It was on the brink of extinction, wiped out by cats on several of its original island homes, and by loss of its scrub habitat.

In a bold rescue operation in 1977 the entire population was transferred to the neighboring Mangere Island. There, free from predators, and with scrub vegetation in which to live, the population has increased to about 30. The black robin is not safe yet, but without human intervention it would certainly have perished.

CAPTIVE BREEDING
Breeding animals in captivity so that they can later be released into their natural habitat can provide another means of preserving endangered species. The Nene or Hawaiian goose is one that was saved in this way.

When Hawaii was discovered by Europeans in the 18th century there were probably more than 25,000 native geese on the islands. Their numbers rapidly dwindled however, and by the middle of the 20th century there were no more than three dozen left. In a dramatic attempt to prevent them slipping into extinction, several birds were shipped to England, to Sir Peter Scott's Wildfowl Trust in Gloucestershire. The gamble paid off. The birds bred well in their new home and several pairs were shipped back to Hawaii. In 1962 the population had increased to over 150, in 1970 it was nearly 300 and still rising: the Hawaiian goose was out of danger.

A LUCKY CHANCE
Mammals too have benefited from captive breeding. The very rare Père David's deer has never been seen in

the wild by modern zoologists. It was discovered in the Imperial Hunting Park in Beijing (Peking) in 1865 by the French naturalist Père (Father Armand) David. A few years later the British ambassador in China shipped a pair of the deer to London Zoo. A few more later went to other major European zoos. This turned out to be most fortunate. In 1894 the Huang-ho River flooded the hunting park and most of the escaping deer were killed by local people for food. Four years later the Duke of Bedford bought up all the deer in European zoos and established a herd at Woburn Abbey. That small herd saved the species. There are now several hundred Père David's deer in the world's zoos, and some have been returned to China.

BREEDING THE CHEETAH

Although some animals breed quite easily in zoos and wildlife parks, many others do not. Some need just the right seasonal weather, others need to be in a large group of their own kind, or be in charge of their own chosen territory. Many different factors control just when, and how successfully,

▲ Captive-bred Indian gavials are wrapped in protective padding before being transported to rivers for release into the wild. The hatchlings are reared until they are 4ft long – by which time they are safe from predators.

▼ In African game parks it is necessary to cull elephants to keep the population at a level the food supply can support. To maintain the well-being of the herd, old or sick animals, and surplus males, are usually selected for the cull.

some species will reproduce, and it is often difficult to provide these conditions in captivity.

Some of these problems were highlighted in the 1970s when London Zoo and the National Zoological Gardens of South Africa tried to breed cheetahs to boost the declining population in Africa. Cheetahs are naturally solitary animals. Males and females normally come together only in order to mate, and so in captivity they need to be kept apart until the female is in breeding condition. Males and females raised together in a zoo show little interest in each other.

There were other problems too. Many male cheetahs are not very fertile. (This was the case with 11 out of 20 males in one captive breeding colony). The dominant males in a captive group will be the ones that mate with the females – but there is no guarantee that the dominant males will also be the most fertile. Also, female cheetahs are not very attentive mothers, and in captivity they often neglect or even kill their young. In the wild they move their cubs frequently from one place to another, but in a zoo enclosure they have less space and often become very nervous. Finally, even when the captive-bred

cheetahs are returned to the wild they often fare badly. They can hunt well enough, but they are not very good at establishing their own territories and are often driven out by wild cheetahs.

PROGRAMED BEHAVIOR

The Snow goose breeds in the high Arctic regions where the summer breeding season may last for only a few weeks. The adult birds know when to breed because they are programed to respond to the length of the day in the far north.

When Snow geese are bred in zoos in temperate regions of Europe or America, they breed successfully – but the young become "programed" to the longer day-length of those regions. The birds are perfectly normal in every way, except that this "programing" makes it impossible for them to be released into their true habitat with any chance of surviving and breeding with their wild relatives.

THE CONDOR'S LAST RESORT

Many scientists believe that taking animals from the wild in order to breed them in zoos should be done only as a last resort, when the alternative is almost certain extinction. That was certainly the case with the California condor. In 1987, in the Sierra Nevada hills of California, American conservationists captured the last three wild condors and took them into the safety of American zoos.

All twenty-seven known California condors are now in captivity. The only hope for the survival of this New World vulture now rests on the captive breeding programme. The last three wild birds were all males, and their addition to the captive gene pool could make a huge difference to the success of the breeding programme.

Eventually it may be possible to release condors into the wild once again, but huge areas of land, free from human interference, will have to be set aside for them.

ZOOS AND SPECIES SURVIVAL

A handful of individuals in zoos around the world may keep a species alive, but it is a completely artificial existence. To survive in the wild, or even be safe in captivity, a much larger population is needed. A small population can appear quite stable for several generations, yet can suddenly crash through disease or some genetic problem, or even through some physical disaster such as flood or fire. Some experts believe that to give any degree of security, at least 500 individuals are needed.

In the past, most zoos tried to collect one or two examples of as many different species as possible. They functioned mainly as zoological collections – living catalogs of all animal and plant life. Now the trend has changed, and many zoos focus on maintaining larger groups of a smaller number of species.

Even so, in order to breed some animals – especially large species such as rhinos, gorillas, tigers and pandas – selected breeding males must often be transported from country to country to be mated with receptive females. This too has its dangers: transporting large animals is a hazardous business,

▲ The Arabian oryx became extinct in the wild in 1972, but survives in several zoos around the world. When a herd was reintroduced to its original habitat in Oman the captive-bred animals could not cope with the true desert.

and the authorities must always be aware of the danger of transporting rabies, foot-and-mouth disease and other infections.

THE FROZEN ZOO

One way of avoiding the colossal expense and the considerable risks involved in shipping live rare animals around the world would be to transport just their eggs and sperm. Artificial insemination is now widespread in cattle farming: in fact most domestic cows today are fertilized in this manner.

With some species it is also possible to store sperm for long periods by freezing them and storing them at very low temperatures. This is possible with the sperm of humans, bulls and a number of other species, but so far it has not been successful with the sperm of pigs and many of the antelopes.

In a number of species it is now possible to fertilize eggs in the laboratory and then store the living embryos in liquid nitrogen, ready to be thawed and implanted into the womb of a receptive female whenever and wherever a new generation of the species is needed.

If the remaining technical and biological problems could be overcome, embryos of countless rare species, and the precious genes they hold, could be stored indefinitely in frozen zoos as a "back-up" to captive breeding and as a long-term insurance policy against their complete loss.

SEED BANKS FOR THE FUTURE
Throughout the world there are now many collections of plant seeds held in cold storage. Most contain seeds of crop plant varieties that are no longer popular but might come back into fashion again one day.

The collection held at the Royal Botanic Gardens at Kew is rather different. It is one of the very few "seed banks" devoted to seeds of wild plants from all over the world. This unique store of plant genetic material contains seeds of more than 5,000 species, variants and sub-species.

The importance of such seed banks becomes clear if we remember how few plant and animal species currently provide most of our food. If anything were to happen to our existing food crops – a dangerous new insect pest or disease for example, or a cultivation problem caused by climatic changes brought on by the greenhouse effect (page 72) – we may have to call on the wild relatives of our crop plants for new genetic material to "build" new and more suitable food plants.

To create a seed bank takes great skill. Plants growing in the wild are very variable, so samples of all the variants should be kept. The seeds must be picked at exactly the right time, just before they fall from the plant, and they must be cleaned of any molds or insects' eggs before being cooled and dried. In most species the moisture content is reduced to 5 percent before the seed is placed in the storage room at −4°F. Most seeds, if well prepared, will last like this for tens if not hundreds of years. (During the Second World War the British Museum herbarium in London was damaged by bombs, and a batch of 147-year-old tree seeds promptly started to germinate when they were soaked by fire-hoses!)

▼The survival of all the great apes now depends heavily on captive breeding. However, many zoo specimens were caught over 20 years ago and now show signs of infertility, all of which makes the zoo keepers' task even more difficult.

▲The survival of rare species can often be helped by using foster mothers. Here a Mountain zebra embryo was implanted into the womb of a pony mare who carried the developing embryo to full term, producing this healthy zebra foal.

GLOSSARY

Acid rain Rain (and snow) that has been made acidic by the chemical pollution of the air by industrial waste gases and car exhaust gas.

Adaptation Any special modification of an animal's body or behavior that suits it to a particular habitat or life-style.

Aerial Associated with the air. Birds such as swifts, which catch all their food on the wing, are called "aerial hunters."

Aggression Any form of behavior in which one animal attacks or threatens another animal.

Aquatic Living in or near the water. See also **Marine**.

Arid Dry; usually used to refer to the lands on the margins of deserts.

Biological control The use of one species (usually a predator or parasite) to control the numbers of another (usually an agricultural pest).

Brood The group of young birds raised in a single breeding cycle.

Browsing Feeding on vegetation above ground level, e.g. the leaves of shrubs, bushes and trees.

Camouflage Colors and patterns on an animal's coat, scales or plumage that enable it to blend in with its surroundings.

Carnivore An animal that feeds mainly on the flesh of other animals.

Carrion Meat from an animal that is already dead before it is found by the flesh-eater.

Cereals A general term for grain crops produced by various members of the grass family e.g. wheat, rice.

Climax The stage in a succession of vegetation types at which no further change takes place unless the physical conditions alter.

Clutch A group of eggs laid during a single breeding cycle.

Colonize Take over a new habitat.

Colony A large group of animals of the same species that lives together all the time (e.g. ants, bees) or that gathers together during the breeding season (e.g. seals, gannets).

Competition A contest between individual animals for mates or food, or between different species over food resources, living space etc.

Coniferous Bearing cones. Coniferous forest is forest consisting mainly of cone-bearing trees such as pines.

Conservation Any actions that are designed to protect and preserve living plants and animals, or their habitats, or the environment in general.

Crustacean Small marine and freshwater animals with hard outer shells. The group includes crabs, shrimps, lobsters, krill and copepods.

Cultivation The management of an ecosystem for the specific purpose of producing food for humans or for their domestic livestock.

Deciduous Describes trees that shed their leaves in autumn or the fall and grow new foliage in spring. Deciduous forest is typically made up of mixed deciduous trees such as oak, ash, sycamore and beech.

Decomposer An organism that obtains all its energy requirements from the dead remains of other organisms, and in doing so breaks down the dead remains so that essential nutrients can be reused in the environment.

Diurnal Active during the day.

Ecology The study of animals and plants in relation to their physical environment and the other animals and plants around them.

Ecosystem A unit which includes all the living organisms and the non-living material within a defined area.

Endangered species One whose numbers have fallen so low that it is in danger of becoming extinct.

Environment The physical surroundings in which a particular species lives, or, more generally the world about us.

Extinction The point at which a species disappears completely, either from one particular area (local extinction) or from the whole world (complete, global extinction).

Fledging The process of growing feathers. The fledging period is the time between hatching and starting to fly. A "fledgling" is a bird that has just learned to fly.

Fodder Food for domestic livestock such as cattle, sheep and goats.

Foraging Moving about in search of food.

Genes The basic units of information, in chemical code, by which physical characteristics are passed from one generation to the next. Combinations of genes control an animal's size, its color, its speed, and so on.

Grazing Feeding on vegetation at ground level.

Greenhouse effect The effect of accumulation of gases such as carbon dioxide in the atmosphere which prevents infra-red radiation escaping from the Earth and so causes global warming.

Habitat The type of surroundings in which an animal lives, for example forest, desert, freshwater lake.

Hatching The process of emerging from the egg; hence hatchling, a bird or insect that has just hatched.

Herbivore An animal that feeds mainly or entirely on plant food.

Insectivore An animal that lives mainly on insects.

Larva (plural larvae) The grub or caterpillar of an insect.

Mammals Animals whose females have mammary glands, which produce milk on which they feed their young.

Marine Associated with the sea, or living in the sea.

Microbes Organisms of microscopic or ultramicroscopic size, such as bacteria, some fungi, and viruses.

Migration The periodic movement of animals from one area to another, sometimes locally but more usually over long distances. Migration is typically in response to seasonal changes in climate and food supply e.g. between summer breeding grounds in the north and warmer wintering areas farther south.

Nestling A young bird that is still in the nest and dependent on its parents.

Nocturnal Active at night rather than during the day.

Omnivore An animal that has a varied diet that includes both animal and plant food.

Parasite An organism (animal or plant) that is totally dependent on another organism (the host) for its nutrient supply.

Photosynthesis The manufacture of sugars, by green plants, from carbon dioxide and water, using sunlight as the source of energy and the green leaf pigment chlorophyll to harness that energy.

Pollution The spoiling or disruption of natural ecosystems or habitats by waste gases, liquids and solid waste produced by human societies.

Population A separate group of animals all of the same species.

Population cycle The fairly regular variation in the numbers of some species, such as voles and lemmings. Lemming populations reach very large numbers every 4 years or so. Mass migrations follow as the animals seek food, and large numbers often perish in attempting journeys that are beyond their ability. Predator populations often rise and fall in unison with those of their prey.

Predator An animal that hunts and kills other animals (its prey) as its main, or only, source of food.

Prey The animals that are hunted by a predator. The word is also a verb, so an eagle is said to prey on small animals such as rabbits.

Rain forest Tropical, subtropical and temperate forests in which rainfall is plentiful and spread throughout the year with no dry season.

Rodents Small animals of the rat, mouse and squirrel group.

Savannah Tropical grassland, usually with scattered trees; typically the wooded grasslands of Africa.

Scavenger An animal that lives mainly on the left-overs of other animals that have either killed or collected food.

Species A distinct animal type that breeds with others of its own kind and produces young that in turn will mature and produce young of their own.

Territory An area in which an animal or group of animals lives, hunts or breeds, and which is usually defended against any intruders.

Toxic Poisonous.

Tropics Strictly, the region between the Tropic of Cancer and the Tropic of Capricorn, 23½° north and south of the Equator. More generally used for the warm, moist zone lying at either side of the Equator.

INDEX

Page numbers in normal "roman" type indicate text entries. **Bold** numbers refer to captions to illustrations. Many of the entries refer to general animal types and not to individual species. Where the text relates to a species, the Latin name of that animal is given in brackets after the common name.

alligators 39, 44
ants 17, **28**, 30
   (eg *Pseudomyrmex*) **28**
   (eg *Crematogaster*) **28**
aphids **77**
avocets **36**

baboons 23
badger (*Meles m eles*) 17
banteng (*Bos banteng*) 84
barnacles 36
bats
   fruit-eating 30
bees 9
beetle
   Carpet (*Anthrenus verbasci*) **79**
   Colorado (*Leptinotarsa decemlineata*) 76, **79**
   Death watch (*Xestobium rufovillosum*) **79**
   Furniture (*Anobium punctatum*) **79**
   Mint leaf (*Chrysolina menthasthri*) **79**
   Woodworm *see* beetle, Furniture
beetles 17, 30, **77**, 78
   bark 17
   carpet 78
   carrion 17
   museum 78
   sexton (*Necrophorus* species) **19**
   woodworm 78
   see also beetle
birds 16, 23, 24, 30, **38**, 40, 60, **61**
   flightless 41
   of prey 24
   wading **36**
bluebottle (*Calliphora vomitoria*) **79**
bongo (*Tragelaphus euryceros*) 36
botfly
   Human (*Dermatobia hominis*) 77

buffalos 36, 57
bunting
   Snow (*Plectrophenax nivalis*) 18, 20
buntings 18
   see also bunting
bush babies 36
bushbuck (*Tragelaphus scriptus*) 36
butterflies 45
   cabbage 22
   see also butterfly
butterfly
   Orange-tip (*Anthocharis cardamines*) **86**
buzzard
   Rough-legged (*Buteo lagopus*) 18

carnivores 14, 22
cat (*Felis catus*) 19, 20, 23
caterpillars 12, 19, **28**
cattle 10, 14, 84, 88
centipedes
   (eg *Lithobius forficatus*) **19**
chamois (*Rupicapra rupicapra*) 36
cheetah (*Acinonyx jubatus*) 87
chough
   Red-billed (*Pyrrhocorax pyrrhocorax*) 81
cockroaches 78
   (eg *Ectobius lapponicus*) **19**
condor
   California (*Gymmogyps californianus*) 88
crab
   Hermit (*Eupagurus bernhardus*) 31
crabs **8**, 36
crocodile
   Orinoco (*Crocodylus intermedius*) 44
crocodiles 39, **44**
   see also crocodile
crossbills 26
crow
   Carrion (*Corvus corone*) 17

deer 9, 16
   Père David's (*Elaphurus davidiensis*) 86
   Red (*Cervus elaphus*) **14**
dinosaurs 40
dodo 41
dolphins 9
dormouse
   Common (*Muscardinus avellanoarius*) **31**
ducks **36**

eagle
   Bald (*Haliaetus leucocephalus*) 63
   Golden (*Aquila chrysaetos*) 26, 42
   Harpy (*Harpia harpyia*) 56
   Wedge-tailed (*Aquila audax*) 42
   White-tailed (*Haliaeetus albicilla*) 42, 86
eagles 36
   see also eagle
earthworms 17
   (eg *Allolobophora turgida*) **19**
egret
   Cattle (*Bubulcus ibis*) 11
   Chinese (*Egretta eulophotes*) 42
egrets **20**
   see also egret
eland (*Taurotragus oryx*) 81
elephant
   African (*Loxodonta africana*) 82
elephants 12, 14, 30, 44, 82, **87**
   see also elephant

fairy-wren
   Black-backed (*Malurus melanotus*) **40**
falcon
   Peregrine (*Falco peregrinus*) **23**, 63
fantail
   Gray (*Rhipidura fuliginosa*) **40**
fish 38
   cleaner 31
   piranha 12
flamingoes **36**
fleas 9
flies 78
   carrion 29
   yellow dung **31**
flycatcher
   African paradise (*Terpsiphone viridis*) **40**
fox
   Arctic (*Alopex lagopus*) 18
foxes 16, 17
   see also fox
frogs 32

gaur (*Bos gaurus*) 84
gavial (*Gavialis gangeticus*) 87
giraffe (*Giraffa camelopardalis*) 84
goats 23, **28**
goose

   Hawaiian (*Branta sandvicensis*) 86
   Snow (*Anser caerulescens*) 88
gorilla (*Gorilla gorilla*) **10**
grasshoppers 26, **83**
grazers 24
grebe (*Podiceps species*) 63
greenfly 19
groupers 31
grouse 26

hare
   Snowshoe (*Lepus americanus*) **27**
hares 26
   see also hare
hawkmoth
   Hummingbird (*Xanthopan morgani praedicta*) 30
hawks 12, 16, 36
hedgehog (*Erinaceus europaeus*) 20
herbivores 14, **14**, 22
heron (*Ardea species*) 63
hogs
   forest 36
horse (*Equus caballus*) 14
hoverflies
   (eg *Chrysotoxum cautum*) **19**
humans (*Homo sapiens*) 7, 23, 40, **49**, 88
hummingbirds 28, 30
hyraxes 36, **36**

ibex (*Capra ibex*) 36
insect
   Cottony cushion scale (*Icerya purchasi*) 78
insects 16, 17, **38**, **39**, 77, **77**
   wood-boring 22
   see also insect

jaguar (*Panthera onca*) 56

kestrel (*Falco tinnunculus*) 22, 46
kouprey (*bos sauveli*) 84

ladybirds 19, 78
lemming
   Arctic (*Dicrostonyx torquatus*) 18
leopard
   Clouded (*Neofelis nebulosa*) 44

Snow (*Panthera uncia*) 44
leopards 36, 46
see also leopard
lice 9
limpets 36
linnet (*Acanthis cannabina*) 26
lizards 32
locust
   Australian plague (*Chortoicetes terminifera*) **77**
   Desert (*Schistocerca gregaria*) 26, **26**
locusts **77**
see also locust
logrunner
   Northern (*Orthonyx spaldingi*) **40**
lynx (*Felis lynx*) **27**

magpies 17
mammals 30, **38**, 40
mice 16, 32, 77
millipede
   Flat-backed (*Polydesmus complanatus*) **19**
millipedes 17
   woodland (*Cylindroiulus* species) **19**
see also millipede
mites 17, **31**
moa 41
mole (*Talpa europea*) 22
monkeys 36
mosquito (*Anopheles* species) 78
   Banded (*theobaldia annulata*) **79**
moth
   Bordered white (*Bupalus pinaria*) **79**
   Case-bearing clothes (*Tinea pellionella*) **79**
   Spruce bud 78
moths 78
see also moth

nene *see* goose, Hawaiian
nyala (*Tragelaphus angasi*) 80

ocelot (*Felis pardalis*) 44
omnivores 23
oryx
   Arabian (*Oryx leucoryx*) **88**
osprey (*Pandion haliaetus*) 63
otter
   Sea (*Enhydra lutris*) **60**
owl
   Barn (*tyto alba*) 22

Snowy (*Nyctea Scandiaca*) 18

parrotbill
   Gray-headed (*Paradoxornis gularis*) **40**
parrots 42
pelican
   Brown (*Pelecanus occidentalis*) **17**
penguin
   Adelie (*Pygoscelis adeliae*) 63
pigeon
   Wood (*Colomba palumbus*) 77
pigs 23, 88
pikes **21**
plover
   Golden (*Pluvialis apricaria*) **23**
plovers 36
see also plover
possum
   Honey (*Tarsipes rostratus*) 30
ptarmigan 18, 26
python
   Australian carpet (*Morelia spilotes*) **22**

rabbit
   European (*Oryctolagus cuniculus*) **24**
rabbits 9, 12, 26
see also rabbit
rat
   Brown (*Rattus norvegicus*) **22, 76**
rats 23, 77
see also rat
redpoll (*Acanthis flammea*) 26
remoras 30
reptiles 7
rhinoceros
   African white (*Ceratotherium simum*) **42, 83, 85**
   Indian (*Rhinoceros unicornis*) **42**
   Sumatran (*Dicerorhinus sumatrensis*) 30
robin (*Erithacus rubecula*) 18
   Chatham Island black (*Petroica traversi*) 86

sand-hoppers 36
sandpiper
   Purple (*Calidris maritima*) 18

sandpipers 18
see also sandpiper
scallops (*Pecten* species) **8**
scorpion
   False (*Dendrochenes cyrneus*) **19**
sea urchins **8**
seals 60
sharks 31
sheep 14
shrews 22
sicklebill
   White-tipped (*Eutoxeres aquila*) 28
sicklebills 30
see also sicklebill
siskin (*Carduelis spinus*) 26
slugs 17
   (eg *Limax maximus*) **19**
snail
   Black-lipped hedge (*Cepaea nemoralis*) **19**
snails 17, 19
see also snail
snakes 44
   sea 32
spiders 19
squirrels 16
   ground 26
sparrowhawk (*Accipiter nisus*) 63
sparrowhawks 19
see also sparrowhawk
starfish
   Crown of thorns (*Acanthaster planci*) 82, **82**
stilt
   Black-winged (*Himantopus himantopus*) **36**
swallow
   Barn (*Hirundo rustica*) 16
swallows 16
see also swallow
swan
   Mute (*Cygnus olor*) **62**
swordbills 30

tapeworms 9
tapirs 30
termites 78
   (eg *Reticuli fermes*) **19**
thornbill
   Yellow-rumped (*Acanthiza chrysorrhoa*) **40**
thrush
   Song (*Turdus philomelos*) 19
thrushes 40
see also thrush
tiger (*Panthera tigris*) 18

tit
   Blue (*Parus caeruleus*) 19, 20
tits 9
see also tit
turtles
   sea 32

warbler
   sedge (*Acrocephalus schoenobaenus*) 24
warblers 9
see also warbler
wasps 30, **77**
waterbuck (*Kobus ellipsiprymnus*) 80
whales
   pilot **42**
whistler
   Golden (*Pachycephala pectoralis*) **40**
wildebeest (*Connochaetes taurinus*) 10, 24, **24**
wolf
   Gray (*Canis lupus*) 41, **45**
woodlice 17
woodpeckers 22
worms 17

yak (*Bos mutus*) 84

zebra
   Mountain (*Equus zebra*) **89**

FURTHER READING

Ayensu, E. S., Heywood, V. H., Lucas, G. I. and DeFillips, R. A. (1984), *Our Green and Living World: the Wisdom to Save It*, Cambridge University Press, Cambridge.

Barnett, S. A. (1981), *Modern Ethology*, Oxford University Press, Oxford.

Colinvaux P. A (1973), *Introduction to Ecology*, John Wiley, New York

Collinson, A. S. (1977), *Introduction to World Vegetation*, George Allen and Unwin, London.

Cox, C. B. and Moore, P. D. (1985), *Biogeography: An Ecological and Evolutionary Approach* (4th edition), Blackwell Scientific Publications, Oxford.

Dawkins, R. (1977), *The Selfish Gene*, Oxford University Press, Oxford

Ehrlich, P. R., Ehrlich, A. H. and Holdren, J.P. (1977), *Ecoscience: Population, Resources, Environment*. W. H. Freeman, San Francisco.

Harris, D. R. (1980), *Human Ecology in Savanna Environments*, Academic Press, London.

Heywood, V. H. (ed) (1985), *Flowering Plants of the World*, Croom Helm, London.

Hora, B. (ed) (1981), *Oxford Encyclopedia of Trees of the World*, Oxford University Press, Oxford.

Krebs, J. R. and Davies, N. B. (1981), *An Introduction to Behavioural Ecology*, Blackwell Scientific Publications, Oxford.

Manning, A (1979), *An Introduction to Animal Behaviour* (3rd edn), Edward Arnold, London.

McFarland, D. J. (ed) (1981), *The Oxford Companion to Animal Behaviour*, Oxford University Press, Oxford.

Mellanby, K. (1981), *Farming and Wildlife*. Collins, London.

Messent, P. and Broom, D. (1986), *The Encyclopedia of Domestic Animals*, Grolier International.

Miller, T. and Armstrong, P. (1982), *Living in the Environment*, Wadsworth, California.

Moore, D. M. (ed) 1982), *Green Planet: The Story of Plant Life on Earth*, Cambridge University Press, Cambridge.

Moran, J., Morgan, M. Wiersma, J. (1986), *Introduction to Environmental Science* (2nd edn), W.H. Freeman, New York.

Myers, N. (1984), *The Primary Source: Tropical Forests and Our Future*, Norton, New York.

Putman R. J. and Wratten, S. D. (1984), *Principles of Ecology*, Croom Helm, London.

Simon J. I. and Kahn, H. (eds) (1984), *The Resourceful Earth*, Basil Blackwell, Oxford

Trivers, R. (1985), *Social Evolution* Benjamin/Cummings, Menlo Park, California.

Tudge, C. (1988), *The Environment of Life*, Oxford University Press, New York.

Whittaker, R. H. (1975), *Communities and Ecosystems*, Collier Macmillan, London. University Press, Cambridge, Massachusetts.

ACKNOWLEDGMENTS

Picture credits

Key: t top, b bottom, c center, l left, r right.
Abbreviations: AN Agence Nature, ANT Australian Nature Transparencies, BCL Bruce Coleman Ltd, NHPA Natural History Photographic Agency, OSF Oxford Scientific Films. P, Premaphotos. PEP Planet Earth Pictures. SAL Survival Anglia Ltd. SPL Science Photo Library.

4 Tony Morrison. 6-7 NHPA/P. Warnett. 7 Mick Saunders. 8t B. Picton. 8b Biophotos. 9 ANT. 10-11 P. Veit. 12 Denys Ovenden. 13t PEP/Chris Prior. 13b Simon Driver. 14 Denys Ovenden. 15 Fred Winner/Jacana. 16 Oxford Illustrators Ltd. 17 Tony Morrison. 18-19 Denys Ovenden. 20-21 Simon Driver. 21 PEP/K. Cullimore. 22 ANT/G. Fyfe. 23 E. & D. Hosking. 24 The Australian Information Service, London. 25 SAL/J. Pearson. 26 Zefa/Tortoli. 27tr Simon Driver. 27bl AN/Lanceau. 27br BCL/S. Kraseman. 28 Premaphotos Wildlife/K. Preston-Mafham. 29 PEP/D. Barrett. 30 PEP/Menhuin. 31t C.A. Henley. 31b Roger Hosking. 32 Marion & Tony Morrison/South American Pictures. 33t G. J. Waugham. 33b OSF. 34l Chris Forsey. 34r Frank Lane Agency/Silvestris. 35 Oxford Illustrators Ltd. 36 A. Bannister. 36-37 Chris Forsey. 38l Simon Driver. 39t Hutchison Library. 39b Robert Harding. 40-41 Norman Arlott. 42 BCL/E. & P. Barker. 43 Zefa/W. Ferchland. 44 K.A. Vliet. 45 Jacana. 46-47 Chris Forsey. 48 Hayward Art Group. 49 Zefa/Kappel Meyer. 50 Hutchison Library. 51 Hayward Art Group. 52 NHPA/P. Johnson. 53 PEP/J. Duncan. 54t USDA. 54b Hayward Art Group. 55 Hutchison Library/John Hatt. 56 Hutchison Library. 57t Robert Harding Associates. 57b PEP/R. Matthews. 58 Tony Morrison. 59 ANT/R. & D. Keller. 60-61 Susan Griggs Agency. 61t PEP/W. Williams. 61 inset Hutchison Library/J. Egan. 62l NHPA/M. Leach. 62r Hayward Art Group. 63 Zefa. 64-65 Greenpeace/Van der Veer. 66 ANT/J.R. Brownlie. 67 N. Collins. 68 US Department of Energy/Science Photo Library. 70 Mick Saunders. 71 Bruce Coleman Ltd/Norman Myers. 72 Mick Saunders. 73l Biofotos/H. Angel. 73r PEP/C. Howes. 74l PEP/D. Gill. 75 NASA/Science Photo Library. 76 NHPA/S. Dalton. 77tl Holt Studios. 77bl P. 77r ANT. 78-79 Richard Lewington. 80 NHPA/D. Woodfall 80-81 Doug Wechsler. 81 NHPA/P. Johnson. 82 PEP/L. Pitkin. 83t Swift Picture Library/N. Dennis. 83b Fermilab. 84,85tr 85bl National Park Board, Pretoria. 86 SAL/D. Green. 87t Dr H.R. Bustard. 87b OSF/G.I. Bernard. 88 NHPA/D.Dick. 89t Zoological Society of London. 89b Ardea/K. Fink.